Tomorrow's Energy

D0762690

Tomorrow's Energy

Hydrogen, Fuel Cells, and the Prospects for a Cleaner Planet

Revised and expanded edition
Peter Hoffmann

The MIT Press
Cambridge, Massachusetts
London, England

665.81
H0f

MIT Press books may be purchased at special quantity discounts for business or sales promotional use. For information, please email special_sales@mitpress.mit.edu or write to Special Sales Department, The MIT Press, 55 Hayward Street, Cambridge, MA 02142.

This book was set in Sabon by Toppan Best-set Premedia Limited. Printed and bound in the United States of America.

Library of Congress Cataloging-in-Publication Data

Hoffmann, Peter, 1935–
Tomorrow's energy : hydrogen, fuel cells, and the prospects for a cleaner planet/
Peter Hoffmann.—Rev. and expanded ed.
    p.   cm.
Includes bibliographical references and index.
ISBN 978-0-262-51695-2 (pbk. : alk. paper)
1. Hydrogen as fuel. I. Title.
TP359.H8H633   2012
665.8'1—dc23

                                                                  2011030564

10 9 8 7 6 5 4 3 2 1

Contents

Foreword

Senator Byron L. Dorgan

While policymakers in the United States struggle with the question of how to become less dependent on foreign oil and at the same time how to reduce greenhouse gas emissions, Peter Hoffmann once again reminds us about a solution that is all around us. It is hydrogen.

In 2001, Hoffmann wrote a book, *Tomorrow's Energy: Hydrogen, Fuel Cells, and the Prospects for a Cleaner Planet*, making the case for the obvious advantages of hydrogen and fuel cells.

Hydrogen is our most abundant carrier of energy, analogous to and interchangeable with electricity. The use of hydrogen and fuel cells to power our vehicle fleet would give us access to a nearly inexhaustible energy medium that is nonpolluting. The exhaust coming out the tailpipe of a vehicle powered by hydrogen is nothing more than water vapor. Pursuing the employment of hydrogen and fuel cells as a fuel source is a win/win proposition for our country: we both protect the environment and reduce our dependence on foreign oil.

You might ask, if the case is that persuasive for hydrogen and fuel cells, why are we not engaged in a crash program to harness the widespread use of this fuel and technology?

The answer is complicated, but in short, the U.S. Department of Energy, after having been a strong advocate of a hydrogen future for more than a decade, has come to believe that commercial-scale use of hydrogen fuel cells is too far out into the future to be a reasonable, viable solution. Although that conclusion is not accurate, the department has severely curtailed, and in some cases eliminated, hydrogen fuel cell research in the past couple of years.

Hoffmann's new book, the revised and expanded edition of *Tomorrow's Energy: Hydrogen, Fuel Cells, and the Prospect for a Cleaner*

Planet, tells us how and why the DOE's position on hydrogen is just flat wrong. The fact is, automobile companies are working hard right now to bring commercial vehicles to the market powered by hydrogen fuel cells. In addition, fuel cells running on hydrogen as well as on other fuels such as natural gas or biofuel gases from wastewater treatment plants, for example, are now being sold commercially and are in fact being installed and used by the thousands, in buildings, forklifts, and much more in many parts of the world.

Around the time of publication of Hoffmann's 2001 book, both President Clinton and President George W. Bush saw great promise in developing hydrogen fuel cells. President Bush announced a major research push to commercialize this clean energy technology. And, frankly, a lot of progress has been made in both public and private sector research, as Hoffmann describes in this new book.

However, things changed when the Obama administration, in its first budget, decided to discontinue most of the federally sponsored hydrogen research. Energy Secretary Chu said that because research dollars are scarce, they should be used for technologies that are more "near term" than hydrogen fuel cells.

I felt the administration was misguided, and as Chairman of the Appropriations Subcommittee that funded the Department of Energy, I added back the funding that had been deleted from the president's budget. That add back continued the federal research for another year, but the administration has not changed its views, and funding for hydrogenand fuel cell research remains in trouble.

The case for change in our energy policy is clear. And Hoffmann's new work describes why continuing working toward a hydrogen future is in our national interest.

Our country has only 3 percent of the world's oil reserves, we are 5 percent of the world's population, we produce 10 percent of the world's oil, and we use over 20 percent of the world's oil. In addition, about 50 percent of the oil we use comes from outside of our country, and 70 percent of our oil is used in transportation. All of this makes a strong case that if we are going to improve our energy security, we need to find ways to power our vehicles with something other than oil. The electrification of the vehicle fleet is the obvious direction. Included in that is a move to hydrogen fuel cell vehicles that are driven by electricity. Hybrids, electric, hydrogen fuel cell vehicles—they are all part of the mix.

In some circumstances, new technology will get to the marketplace through the efforts of the private sector without help from government policy. But in the case of converting to a hydrogen fuel cell future, the issues of production, storage, distribution, and related matters require that public policy leads the way.

Those of us who believe that our country will benefit from a hydro-gen/fuel cell future are indebted to Peter Hoffmann for his work over many years in support of this opportunity for America's future. This second edition will be a major contribution to the debate about our energy future.

Senator Dorgan, Democrat from North Dakota, served in the United States Senate from 1992 to 2011 when he retired. Since then he has continued to work on energy issues as a Senior Fellow at the Bipartisan Policy Center in Washington, D.C., where he is cochairman, with former Senator Trent Lott (R-Miss.), of its energy project.

Preface and Acknowledgments

This revised and expanded edition of *Tomorrow's Energy* continues to report on global developments and trends—both advances and setbacks— of the evolving hydrogen energy economy in the ten years since the first edition was published in 2001. It traces what has happened in this field roughly since the year 2000 while retaining much of the earlier material, including the discovery and early history of hydrogen and the late-twentieth-century efforts to harness hydrogen as transportation fuel for road vehicles, airplanes, and other uses. Today, more than ever before, the evolution of an energy economy based on nonpolluting, zero-emission hydrogen as fuel is both inevitable and imperative in light of worldwide concerns over energy security and global warming caused by carbon dioxide.

This edition covers the major aspects of production, storage, transportation, use as a utility fuel, and safety. It also touches on some of the recent political controversies in the United States surrounding hydrogen— for example, the views of Secretary of Energy Steven Chu versus those of hydrogen and fuel cell supporters and scientists in the United States and abroad, the findings of think tanks and major consultants, the National Research Council, and practically everybody else, including major carmakers and developers of European and Japanese hydrogen infrastructure plans, for example. It also introduces some concepts not widely known, such as hydricity—the essential interchangeability of electricity and hydrogen—and prospects for both fuel cell–powered small aircraft as well as hydrogen-fueled hypersonic airplanes. Finally, it gives a glimpse of what some thinkers and practitioners guess may happen in coming decades.

As in the original book, many people contributed information, data, advice, and, last but certainly not least, moral support when I despaired that it was just all too much to go through again. The most important person was Sarah Hoffmann—my wife, collaborator, and colleague for more than four decades, sharp-eyed editor, co-translator, text improver, and, when needed, heckler to get me up off the floor (make that "couch") and back to the computer—and Giotto, my brilliantly perceptive shepherd-chow-lab mix who takes me for walks when I need to air out my atrophying little gray cells.

There were many others, and I will probably miss a few. I am thankful to all of them more than I can say. They include, in no particular order, David Hart, Paul Hesse, Sandy Thomas, Jim Joosten, Michael Graetzel, Darlene Steward, Margaret Mann, Marc Melaina, Rick Farmer, Sunita Satyapal, Sam Atwood, Lawrence Abramson, Timothy Volk, Steve Shi, Jan Tuchman, Scott Lewis, Cesare Marchetti, Carol Worster, John O'M. Bockris, Melanie Cecotti, Matthias Brock, Caroline Fife, Addison Bain, Michael Bernstein, Ben Mehta, John Appleby, Chris Borroni-Bird, Josh Lieberman, Bill Craven, Lawrence Smart, Jeff Serfass, Patrick Serfass, Kyle Gibeault, Sandy Bartlett, John Turner, Paul Grabowski, Joan Ogden, Dimitri Stanich, Alan Lloyd, Kristen Nelson, Linda Church-Ciocci, Brian Murphy, Jon Bjorn Skulason, Maria Maack, Hjalmar Arnason, Garret Drexler, Elizabeth Salerno, Sarah Howell, Luis Vega, Barbara Heydorn, Peter Lehman, Kevin Harris, Byron McCormick, George Hansen, Henry Wong, Tobias Hahn, Jennifer Gangi, Reinhold Wurster, Kerry-Ann Adamson, Kurt Goddard, Peg Hashem, Steve Gitlin, Johan Steelant, Dave Ercegovic, Mark Hemsell, Kolja Seeckt, Andreas Westenberger, Agata Godula-Jopek, Barton Smith, Monterey Gardiner, H. T. Everett, Ulrich Buenger, Christoph Stiller, Lars Sjunneson, Walt Pyle, Robert Rose, John Hunter, Pin-Ching Maness, Mike Seibert, Paul Horowitz, James Riordon, Randy Dey, Ulrich Schmidtchen, William Vincent, Peter Kolp, Nebosja Nakicenovic, Nick Easley, Hirohisa Uchida, Nejat Veziroglu, Juergen Garche, Michael Oppenheimer, Dan Sperling, Kazukiyo Okano, Sven Geitmann, Mikael Sloth, Monique Dunn, Brian Tian, German Lewizki, and Saeed Moghaddam.

A special thank you goes to Nejat Veziroglu, founder and president of the International Association for Hydrogen Energy, for financial assistance in this project.

And finally, I thank Clay Morgan, my lifeline and contact at MIT Press who patiently has put up with me for both editions of this book, his assistant Laura Callen, and my MIT Press editors, Sandra Minkkinen and Beverly Miller, who smoothed out wrinkles and bumps in my scribbles. Without them, this would have been mission impossible.

1

Why Hydrogen? The Grand Picture

There are two prime sources of energy to be harnessed and expended to do work. One is the capital energy-saving and storage account; the other is the energy-income account. The fossil fuels took multimillions of years of complex reduction and conservation, progressing from vegetational impoundment of sun radiation by photosynthesis to deep-well storage of the energy concentrated below the earth's surface. There is a vast overabundance of income energy at more places around the world, at more times to produce billionsfold the energy now employed by man, if he only knew how to store it when it is available, for use when it was not available. There are gargantuan energy-income sources available which do not stay the processes of nature's own conservation of energy within the earth's crust "against a rainy day." These are in water, tidal, wind, and desert-impinging sun radiation power. The exploiters of the fossil fuels, coal and oil, say it costs less to produce and burn the savings account. This is analogous to saying it takes less effort to rob a bank than to do the work which the money deposited in the bank represents. The question is cost to whom? To our great-great-grandchildren who will have no fossil fuels to turn the machines? I find that the ignorant acceptance by world society's presently deputized leaders of the momentarily expedient and the lack of constructive, long-distance thinking—let alone comprehensive thinking—would render dubious the case for humanity's earthian future could we not recognize plausible overriding trends.
—R. Buckminster Fuller, 1969[1]

The big powers are seriously trying to find alternatives to oil by seeking to draw energy from the sun or water. We hope to God they will not succeed quickly because our position in that case will be painful.
—Sheikh Ahmad Zaki Yamani, oil minister of Saudi Arabia, 1976[2]

Hydrogen as fuel? It's still Buck Rogers stuff.
—Energy expert, Bonn, February 1980

Ballard Power and United Technologies are leading pioneers in developing fuel cells that are so clean. Their only exhaust is distilled water. Right now, Ballard is working with Chrysler, Mercedes-Benz and Toyota to introduce fuel cells into new cars.
—President Bill Clinton, 1997[3]

In the twenty-first century hydrogen might become an energy carrier of importance comparable to electricity. This is a very important mid- to long-range research area.
—President's Committee of Advisors on Science and Technology, 1997[4]

Now analysts say that natural gas, lighter still in carbon, may be entering its heyday, and that the day of hydrogen—providing a fuel with no carbon at all, by definition—may at last be about to dawn.
—*New York Times*, 1999[5]

We asked ourselves, is it likely in the next 10 or 15 or 20 years that we will convert to a hydrogen car economy? The answer, we felt, was no.
—Secretary of Energy Steven Chu, 2009[6]

This study shows that FCEVs [Fuel Cell Electric Vehicles] are technologically ready and can be produced at much lower cost for an early commercial market over the next five years. The next logical step is therefore to develop a comprehensive and co-ordinated EU market launch plan study for the deployment of FCEVs and hydrogen infrastructure in Europe.
—2010 McKinsey study[7]

These quotations give some idea as to what this book is all about: hydrogen as a nonpolluting, renewable form of energy. Hydrogen—an invisible, tasteless, colorless gas—is the most abundant element in the universe. It is the fuel of stars and galaxies. Highly reactive, it is essential in innumerable chemical and biological processes. It is an energetic yet (by definition) nonpolluting fuel.[8]

Even before Buckminster Fuller's observations, many people had been calling for the use of nature's "current energy account" (solar power in its various manifestations) as an alternative to robbing the world's energy "savings account" (coal, oil, gas). As Fuller pointed out, the problem has been to a large extent not only how to collect this essentially free energy but how to store it. Tapping into solar energy for purposes other than basic solar heating usually means producing electricity. But electricity has to be consumed the instant it is produced because it is difficult to store in large quantities. Hydrogen, a storable gas, solves that problem.

In past decades, efforts to harness renewable energies were driven partly by idealism but more by concerns about energy security—that is, fears that the world's petroleum resources will eventually dry up and about the increasing vulnerability of the long supply lines from the politically unstable Middle East. But as the twentieth century drew to its close and the twenty-first century arrived, environmental concerns became much stronger, to the point that today they dominate much of the national and global political discourse, driving the world toward renewable, alternative forms of energy. Curbing and eventually doing away with pollution has become a global universal concern. Dying forests in Europe and acid rain everywhere were among the initial calls for the need to curb sulfur, nitrogen oxides, hydrofluorocarbons, perfluorocarbons, particulate emissions, and other pollutants. Today it is clear to almost everyone in the world that the very process of combusting fossil fuels, the interaction of carbon in hydrocarbon fuels with the air's oxygen, and the consequent release into and accumulation in the atmosphere of carbon dioxide and other climate-changing gases far above preindustrial levels is raising the world's temperature—the infamous greenhouse effect—and threatening to play havoc with the world's climate.

The new world standard is becoming zero emissions from cars and buses, industry, ships, and home furnaces, a standard to which industrialized countries and emerging economies are aspiring with varying degrees of intensity and dedication. To the minds of many, taking the carbon out of hydrocarbons and relying on the "hydro" part—hydrogen—as a zero-emission chemical fuel is the obvious, though technically difficult, way to minimize and, it is hoped, eventually eliminate global warming.

The basics of global warming are roughly as follows. Carbon dioxide (CO_2) is produced by the burning of fossil fuels as well as by nature's carbon cycle. (Humans and animals exhale it into the atmosphere as part of their metabolic process; green plants absorb it and turn it into plant matter.) CO_2, methane, and other gases act like a greenhouse in the atmosphere: They let solar radiation through the atmosphere to heat the earth's surface, but they prevent the reradiation of some of that energy back into space, thus trapping heat. Some heat entrapment is good; otherwise we would have never evolved in the first place, or we would freeze to death. But the more greenhouse gases are swirling around the atmosphere, the more heat is trapped. Because of diminishing forests around the world and consequent decreases in global CO_2 absorption, and (more important) because of increasing burning of fossil fuels in our ever-more-energy-demanding industrial machinery, the atmosphere's CO_2 content has been going up steadily and increasingly steeply since the beginning of the Industrial Revolution.

Aside from other fundamental climate cycles stretching over thousands or tens of thousands of years (such as ice ages, believed to be caused in part by changes in sunspots and therefore beyond human ability to influence), the earth's climate has been reasonably stable for 10,000 years or so. But this equilibrium is being upset by human-made carbon emissions. The question is how much. Opinions, basic assumptions about the future course of the climate and the amount of expected heat increase, closely related assumptions about global economic development, and faith in the complex computer models that attempt to forecast climate developments vary widely even among the majority of experts who believe that our planet is facing an unprecedented crisis.[9]

As more heat is being trapped and as temperatures climb all over the world, the mainstream opinion among the climate experts of the United Nations' Intergovernmental Panel on Climate Change predicts widespread and drastic impacts on ecosystems, water resources, food and fiber production, coastlines, and human health: the polar ice caps will melt, sea levels will rise, large stretches of coastline (including some of the world's biggest cities) will be inundated, and scores of islands in the Pacific may disappear. Agricultural patterns are likely to change, with grain-growing belts migrating northward. The middle to high latitudes may become more productive as plants absorb more available CO_2. The agricultural yields of the tropics and the subtropics are expected to decrease.

The December 2009 Copenhagen Climate Summit was widely regarded as disappointing in its outcome, with little real accomplishment. Nevertheless, the Copenhagen Accord's opening paragraph clearly and starkly states the basic problem and what needs to be done:

We underline that climate change is one of the greatest challenges of our time. We emphasize our strong political will to urgently combat climate change in accordance with the principle of common but differentiated responsibilities and respective capabilities. To achieve the ultimate objective of the Convention to stabilize greenhouse gas concentration in the atmosphere at a level that would prevent dangerous anthropogenic interference with the climate system, we shall, recognizing the scientific view that the increase in global temperature should be below 2 degrees Celsius, on the basis of equity and in the context of sustainable development, enhance our long-term cooperative action to combat climate change. We recognize the critical impacts of climate change and the potential impacts of response measures on countries particularly vulnerable to its adverse effects and stress the need to establish a comprehensive adaptation program including international support.

Ten years earlier, a 1999 study that looked at the generation of ozone in four metropolitan areas (Sacramento, Chicago, St. Louis, and Los Angeles) concluded that a future doubling of global atmospheric CO_2 would likely result in higher daily temperatures, which "dominate the meteorological correlations with high tropospheric ozone concentrations"—in other words, higher temperatures would increase the ozone concentrations.[10] More ozone, in turn, would increase the incidence of premature mortality, hospital admissions for respiratory diseases, and respiratory symptoms, the authors said. But some aspects, especially the relationship between ozone levels and premature mortality, are still subject to ongoing research, one author cautioned. In the case of Los Angeles, doubled CO_2 concentrations were expected to increase the annual average daily maximum temperature from the base case 20.7°C to 24.9°C and the annual average daily minimum from 14.1°C to 18.2°C, the researchers calculated. In Chicago, doubled CO_2 would increase the corresponding maximum from 13.5°C to 19.3°C and the minimum from 3.78°C to 10.0°C. For Los Angeles, a table of anticipated extra health costs for one such warmer future year listed $2.552 billion (in 1990 dollars) for premature mortality, $14.19 million for hospital admissions, and $168,000 for respiratory-symptom-days relative to the same cost categories for a typical recent year. For Chicago, the corresponding numbers were $979 million, $2.38 million, and $28,000.

The other principal form of clean energy, electricity, has two strikes against it. First, it is the minority component in the world's energy production and consumption—chemical energy accounts for almost two-thirds. Second, most electricity is produced by burning fossil fuels—coal, natural gas, and petroleum.

According to the May 2009 edition of the U.S. Energy Department's *International Energy Outlook*, total world "marketed" energy consumption (as opposed to "nonmarketed" energy sources, which the report says continue to play an important role in some developing countries) was 472 quads (quadrillion Btu) and was projected to grow to 552 quads by 2015 and 678 quads in 2030.[11] Most of the growth—73 percent—is projected to occur in countries outside the Organization for Economic Cooperation and Development (OECD), with 15 percent growth in the thirty members of OECD, most of them high-income industrial states. Total electricity production in 2006 was 186.3 quads from all sources (liquids, coal, natural gas, nuclear, and renewables), projected to rise to 231.7 quads in 2015 and 300.3 quads in 2030. Coal was the biggest primary source, with 127.5 quads in 2006, 150.7 quads in 2015, and 190.2 quads in 2030. Oil (renamed "liquids" because it now includes biofuels and liquid fuels produced from natural gas and coal) consumption was 172.4 quads in 2006, expected to rise to 183.3 quads in 2015 and 215.7 quads in 2030. Assuming a low-growth scenario, renewable energy, including hydroelectric power, accounted for only 36.8 quads in 2006, rising to 54.0 quads in 2015 and 74.1 quads in 2030. World nuclear energy consumption stood at 27.8 quads in 2006, rising to 31.9 quads by 2015 and 40.2 quads in 2030. And CO_2 emissions, which amounted to 29 billion metric tons in 2006, are slated to rise to 33.1 billion tons in 2015 and 40.4 billion metric tons in 2030, an increase of 39 percent.

Thus, it is safe to say that in general, we work and play with—and, environmentalists would say, increasingly die from—fossil-fueled chemical energy. Gasoline, diesel fuel, heavy oil, jet-grade kerosene, natural gas, wood, biomass, and coal propel airplanes, cars, trains, and ships; run plants; and heat homes, offices, hospitals, and schools. Hydrogen, also a form of chemical energy, can do all those things, and can do them essentially without polluting.

When burned in an internal-combustion engine (piston, rotary, or gas turbine), hydrogen produces mostly harmless water vapor (plus, admittedly, trace emissions from tiny amounts of engine lubricants that are oxidized in the process, and some polluting nitrogen oxides), meaning that an internal combustion engine, even when operating on hydrogen, is not a zero-emission vehicle.[12] When hydrogen is combusted with atmospheric oxygen in an engine, no CO or CO_2 is emitted, no unburned hydrocarbons, no stench, no smoke, and none of any of the other carbon-bearing, earth-befouling discharges we suffer today.

Hydrogen performs even better in fuel cells: electrochemical engines that electrochemically combine hydrogen and oxygen in a flameless process and produce electricity, heat, and pure, distilled water—the mirror image of electrolysis, in which water is split into hydrogen and oxygen by running a current through it. Unlike internal combustion engines, fuel cells produce no nitrogen oxides at all.[13]

Fuel cells have no moving parts. Nearly silent, they can be as much as two and a half times as efficient as internal combustion engines. As one example, the fuel cell version of a Toyota Highlander SUV has demonstrated three times better fuel economy than its standard gasoline version.[14] Beginning in earnest in the 1980s and 1990s, fuel cells have become widely recognized as a vanguard technology that may launch hydrogen energy on its way to becoming a major environmentally benign, sustainable, renewable component of the world's energy mix for both transportation and stationary applications.

"Hydrogen, H_2, atomic weight 1.00797 . . . is the lightest known substance," reports the *Encyclopedia of Chemistry:*

The spectroscope shows that it is present in the sun, many stars, and nebulae. Our galaxy . . . plus the stars of the Milky Way is presently considered to have been formed 12 to 15 billion years ago from a rotating mass of hydrogen gas which condensed into stars under gravitational forces. This condensation produced high temperatures, giving rise to the fusion reaction converting hydrogen into helium, as presently occurring in the sun, with the evolution of tremendous amounts of radiant thermal energy plus the formation of the heavier elements. Hydrogen gas has long since escaped from the Earth's lower atmosphere but is still present in the atmosphere of several of the planets. In a combined state, hydrogen comprises 11.19 percent of water and is an essential constituent of all acids, hydrocarbons, and vegetable and animal matter. It is present in most organic compounds.[15]

Hydrogen is used in many industries as a chemical raw material, especially in the production of fertilizer, but also in making dyes, drugs, and plastics. It is used in the treatment of oils and fats, as a fuel for welding, to make gasoline from coal, and to produce methanol. In its supercold liquid form, in combination with liquid oxygen, it is a powerful fuel for the space shuttle and other rockets.

Hydrogen is produced commercially in almost a dozen processes. Most of them involve the extraction of the "hydro" part from hydrocarbons. The most widely used, least costly process is steam reforming, in which natural gas is made to react with steam, releasing hydrogen. Water electrolysis, in which water is broken down into hydrogen and oxygen by running an electrical current through it, is used where electricity is cheap and high purity is required.

Hydrogen can be stored as a high-pressure gas or as an integral component in certain alloys known as hydrides, as so-called chemical hydrides, in and on microscopic carbon fibers, and with other only recently developed sophisticated technologies. As a cryogenic liquid fuel, it promises to lead to better, faster, more efficient, environmentally clean airplane designs. Metallic hydrogen, a laboratory curiosity so far, holds eventual, distant promise as an ultra-energetic fuel and also as a zero-resistance electrical conductor in all sorts of electrical and electronic technologies—if it can be made in sufficient industrial quantities and be made stable at higher, near-ambient temperatures.

Since the 1930s, environment-minded scientists, academics, and energy planners (inside and outside government), industrial executives, and even some farsighted politicians have been thinking of and supporting the concept of hydrogen as an almost ideal chemical fuel, energy carrier, and storage medium. As a fuel, it does not pollute. As an energy-storage medium, it would answer Fuller's call for some method "to store [energy] when it is available for use when it is not available."

Hydrogen is not an energy source, a mistake that otherwise sophisticated, well-informed people still make. That is, it is not primary energy like natural gas or crude oil that exists freely in nature. It is an energy carrier—a secondary form of energy that has to be manufactured like electricity, which does not exist freely in usable form either. One fascinating aspect is the complementarity of hydrogen and electricity, of great relevance to the whole notion of fuel cell cars and electrification of the automobile, something that carmakers are now pursuing intensively. Hydrogen can be converted in a fuel cell to electricity, plus some heat and water, via the electrochemical combination with the air's oxygen. Running in reverse—and some experimental two-way devices have been built—water can be split into hydrogen and oxygen with an electric current in an electrolyzer. The late Geoffrey Ballard, founder of the Canadian fuel cell developer Ballard Power Systems, coined the word *hydricity*, for hydrogen and electricity, around the turn of the century to describe this phenomenon.

Hydrogen can be generated from many primary sources—an advantage in itself, since it reduces the chances of creating a hydrogen cartel similar to OPEC. Today hydrogen is made (that is, extracted) mostly from fossil fuels. But efforts to clean up these fuels (to "decarbonize" them, in the jargon of energy strategists) will increase. "To decarbonize" really means to adapt and improve techniques long used in the chemical, petroleum, and natural gas industries to strip out the carbon or CO_2 and store ("sequester") it out of harm's way, leaving hydrogen.

In the future, hydrogen will be made from clean water and clean solar energy, from biomass and biofuels, even from coal —and possibly from cleaner versions of nuclear energy. Since it can be made from both non-renewable and renewable sources, it can be phased into the overall energy structure by whatever method is most convenient and least wrenching to a given country, state, region, or economy. Possibilities are coal gas-ification in the western United States and solar-based electrolysis in deserts in the Middle East or the southwestern United States. Israeli scientists are testing direct solar water splitting, in which the sun's con-centrated heat would break up water molecules into hydrogen and oxygen. Water could be electrolyzed with electricity produced by geo-thermal resources in some areas, and perhaps also from the oldest form of renewable energy: hydropower.

In the simplest terms, the broad outlines of a future "hydrogen economy" run something like this. Clean primary energy—probably solar energy in its many variations; possibly an advanced, environmen-tally more benign version of nuclear energy—would produce electricity, which would be used to split water into hydrogen as fuel, with oxygen as a valuable by-product. Alternatively, heat produced by solar or nuclear power plants would be used to crack water molecules thermochemically in processes now under development. More exotic methods in which hydrogen is produced—from genetically engineered microbes, algae, cel-lulosics, and other biological processes—are likely candidates further down the road.

Hydrogen would be used as an energy-storage medium—as a gas under pressure in large, depleted natural gas fields perhaps, in hydrogen-absorbing alloys (the above-mentioned hydrides), as a cryogenic liquid, or in activated-carbon materials and carbon nanostructures; but also in the form of relatively conventional fuels, such as methanol, which some regard as a superior fuel to carry hydrogen. Hydrogen could fulfill the indispensable storage function of smoothing the daily and seasonal fluc-tuations of solar power and other intermittent energy sources.

Hydrogen could be burned in modified internal combustion engines that ordinarily run on fossil fuels—jets, turbines, four-strokes, two-strokes, Wankels, diesels. This was the vision, conviction, and message of hydrogen's supporters from the 1970s through the mid-1990s. Since then, with sudden and rapid advances in fuel cell technology, the emphasis has shifted dramatically toward fuel cells as the future engines of choice for transportation[16] and also as clean, efficient, decentralized sources of electricity for buildings. Ultimately, fuel cells operating on pure hydrogen would be quintessentially clean, producing no nitrogen oxides and no hydrocarbons. The only stuff coming out an exhaust pipe would be harmless water vapor (steam), which returns to nature's cycle of fog, clouds, rain, snow, groundwater, rivers, lakes, and oceans. That water could then be split again for more fuel.

As a gas, hydrogen can transport energy over long distances, in pipelines, as cheaply as electricity (under some circumstances, perhaps even more efficiently), driving fuel cells or other power-generating machinery at the consumer end to make electricity and water.[17]

As a chemical fuel, hydrogen can be used in a much wider range of energy applications than electricity. For example, it is difficult to envision a large commercial airliner powered by electric motors of any conceivable type, but hydrogen is seen as a promising future fuel for aviation (see chapter 8). In addition, hydrogen does double duty as a chemical raw material in a myriad uses. And unlike other chemical fuels, it does not pollute.

Two major goals of international hydrogen research have been to find economical ways of making the fuel and determine how to store it efficiently onboard a space-constrained car, bus, or truck. During the 1970s and the 1980s, much, if not most, of the hydrogen production research was aimed at splitting large volumes of water molecules. This was perceived as the crucial prerequisite to using hydrogen as a fuel. In the 1990s, the emphasis shifted to making hydrogen energy—not necessarily ultra-pure hydrogen—an industrial and commercial reality. Thus, much more attention has been paid to improving the steam reforming of natural gas. For a while, the efforts of some carmakers to use methanol as a hydrogen carrier for fuel cell vehicles represented another example. It had some ecological appeal because methanol, produced industrially from natural gas, can also be made without major impact on the atmosphere ("carbon dioxide neutral" is the catchphrase) from green plants (biomass) that absorb CO_2 in their growth phase.[18] That appeal has not totally faded away. Methanol is still being promoted by some, and making commercial headway, as an energy source for backup power for small fuel cells for handheld electronics and telecommunications. It is also being promoted for transportation. A third approach was exemplified by the U.S. Department of Energy's logistics-driven strategy of developing, in cooperation with major carmakers, onboard fuel processors that would extract hydrogen from gasoline and other fossil fuels, including methanol. That effort has been discontinued, however, largely because of the added complexity of onboard fuel processing—a "chemical factory under the hood" was the slightly derisive phrase.

In past decades, hydrogen advocates believed that a global hydrogen economy would begin to take shape near the end of the twentieth century and that pure hydrogen would be the universal energy carrier by the middle of the twenty-first century. Hydrogen may not completely attain that lofty status—a lot of other clean energy technologies, promoted by a lot of players, have sprung up in the past two decades—but it is certain to play a much larger role in the decades ahead—directly as a fuel for fuel cells and indirectly as an increasingly large component of carbon-based fuels such as methanol and other conventional fuels. Many see it as an increasingly important complement to electricity. Electricity and electrolysis can break water down into hydrogen and oxygen, and hydrogen recombined with oxygen can produce electricity and water again—the hydricity concept already mentioned. Each will be used in areas where it serves best, and for a long time to come, it will have to compete with, and in fact be dependent on, conventional fossil fuels as its source.

What about nuclear power as a primary energy source for the production of hydrogen? The instinctive short answer from most hydrogen supporters and environmentalists probably is that nuclear power's days have come and gone. As one American antinuclear protester (Claire Greensfelder, coordinator of the Berkeley-based group Plutonium Free Future) famously put it in a CNN interview during the December 1997 Kyoto climate negotiations, "Trying to solve climate change with nuclear power is like trying to cure the plague with a dose of cholera." But that wasn't always so. In fact, in the 1970s, many in the hydrogen community counted on atomic energy as a source of cheap power for splitting the water molecule. As a cosmic energy dance combining the elementary force that heats the sun and the other stars and the elementary building block of all matter, the concept had an almost mystical elegance. But while a nuclear fire burning far away in the cosmos is one thing, building a nuclear reactor in a populated area is quite another—or so it seemed to the increasingly powerful antinuclear forces around the world. In the mid-1970s, orders for new nuclear plants began a sharp decline. And then came Three Mile Island (1979) and Chernobyl (1986). It looked as if those two accidents would be the gravestones of the nuclear age. The debate is not over, though. Some long-term energy thinkers, including some with very good environmental credentials, believe that a second wave of environmentally much more acceptable nuclear power stations may well be inevitable and may become a reality in this century. This notion gained traction in the Bush administration and was embraced by the Obama White House as well, although large doubts remain— especially in the wake of the mid-March 2011 tsunami-caused nuclear plant disaster in Fukushima, Japan.

The 1980s were a bad time for environmentalists and clean energy advocates. In the United States, the Reagan administration was basically apathetic to their long-term planetary concerns, focusing instead on military and geopolitical matters. Interest in clean, renewable energy, including hydrogen, did not pick up again until the early 1990s, when worries over environmental issues were mounting. It is probably impossible to give an exact date when that interest got started again, but as good a landmark as any was the publication of Al Gore's 1992 book, *Earth in the Balance: Ecology and the Human Spirit,* augmented by his 2006 documentary, *An Inconvenient Truth.*[19]

For the international community of hydrogen researchers and supporters, a defining moment came in the spring of 1993, when Japan's government announced its WE-NET (World Energy Network) project. This long-range project to help launch hydrogen as the world's clean energy currency of choice was an outgrowth and a redefinition of Project Sunshine, a national multidimensional alternative energy project begun in 1974. The original announcement said that Project Sunshine was to extend until 2020. It would spend the equivalent of about $2 billion on most aspects of hydrogen energy technology—a level of funding and a truly long-term planning horizon, appropriate to the momentous task of addressing a planetary issue such as global warming, that the governments of Western Europe and North America were neither capable of nor particularly interested in at the time. As it turned out, Japan's annual funding for hydrogen was more modest in these early years than expected in the first rush of enthusiasm, both because WE-NET's planners decided to start slowly and cautiously, first analyzing what was needed, and because Japan's once seemingly unshakable economy suffered severe setbacks in the ensuing years. Still, WE-NET was the world's first major hydrogen-centered response by a major industrial country to the growing concerns about global climate deterioration caused by fossil fuels.

Also in the early 1990s, the threat that CO_2 and other trace gases might heat up our planet excessively began to command much more public attention, perhaps faster in Europe and elsewhere than in the United States. Since the 1992 Rio de Janeiro Earth Summit (which many regarded as ineffectual grandstanding), global warming has been reported, discussed, analyzed, dissected, argued, and fought over in countless news stories, interviews, magazines, op-ed pieces, scholarly and popular books, TV programs, and Internet postings.

In the first decade of the twenty-first century, there were still plenty of global warming deniers, in the U.S. Congress notably Senator James Inhofe (R, Oklahoma) and Congressman Dana Rohrabacher (R, California), and there are many other doubters around the world. Supporters of renewable, alternative, carbon-neutral, zero-emission energy technologies say it is better to be safe than sorry. In the decade up to the year 2000, the business-as-usual course was the one much preferred and vigorously lobbied for by the world's traditional energy industries and their allies, documented exhaustively and persuasively by Ross Gelbspan in his book *The Heat Is On* (1997), but since then, evidence has been growing that big oil, coal, and other industries are in the process of changing their minds—if slowly and in fits and starts.[20] Greenhouse gases exist in tiny fractions in the atmosphere—only parts per million and even per billion. A fear is that a minuscule change in concentrations could trigger big, unanticipated, and possibly traumatic change in the atmosphere. As Alan Lloyd, the secretary of the California Environmental Protection Agency under Governor Arnold Schwarzenegger and one of the pivotal figures on the American hydrogen scene since the 1990s, put it in March 1998, addressing a Society of Automotive Engineers fuel cell workshop in Cambridge, Massachusetts, "Environmental pollution will likely represent the 'cold war' of the next century."

If hydrogen's benefits as a fuel are so great, the average person might reasonably ask why it did not make significant inroads into our energy systems years or even decades ago. There is no single, simple answer to that question; there is instead a complex array of interlocking factors. For one, there was no real use for hydrogen as long as there were ample supplies of oil and natural gas and as long as environmental worries were the concerns of a tiny, near-silent minority. Hydrogen's principal advantage over conventional fuels is that it is emission free. That, by itself, was not thought to merit a society-wide switch to alternatives of any sort. Fossil fuels were cheap, and hydrogen was as much as several times more expensive. Liquid hydrogen, the coldly exotic stuff that powers the space shuttle and experimental BMW sedans today, was a laboratory curiosity four or five decades ago.

Technologically the level of development was such that producing, handling, and storing hydrogen was complex, difficult, and perhaps beyond the abilities of the routine consumer. It still is, although it has been improving rapidly and dramatically to the point where hundreds of hydrogen cars are operating on public roads, many of them driven by average everyday drivers, not trained technicians.

Bringing a technology to maturity takes time. David Hart, a consultant and director of the London- and Lausanne-based sustainable energy consultancy E4tech and Principal Research Fellow at Imperial College London, observed in 2000, "We have only recently become able to operate really well with natural gas." He believed the time was finally at hand when hydrogen would start to make major inroads because of "a confluence of drivers that all point in the same direction—towards hydrogen." The drivers include the requirement for a reduction in CO_2 emissions, appalling urban air quality, legislation dictating zero-emission vehicles, progress in fuel cell technology, a move toward the use of local resources for energy production, the need to store intermittent renewable energy, concerns about fossil fuel resources, and the security of energy supplies. Hart concluded: "There is only one common thread running through these, and that is hydrogen. While other energy carriers can assist in achieving some of these objectives, none of them meet all of the requirements. That is why even the major oil companies see hydrogen as a major part of the energy future."[21]

Automobiles have been around for more than 100 years, yet even the best-engineered examples still break down. Perhaps most important, societal issues have prevented major progress. For one, replacing an entire technologically advanced energy system with something else is a huge undertaking, spanning decades. It is like trying to change the course of a supertanker with kayak paddles. Noted one expert in the 1990s, "The energy system consists of an immense infrastructure, enormous physical and human capital, not only tanks and pipelines and motors, but also people—bankers, auto mechanics, drillers, etc. (and politicians, he might have added), hence it evolves slowly." Phasing in hydrogen requires "innumerable replacements"; substituting fuel cells for internal combustion engines is only one small aspect.

Perhaps the biggest impediment to change for the better is our value system—what we are willing to pay for. By and large, environmental health is not high on the list. As one American expert with experience in both the halls of Congress and hands-on alternative energy research, C. E. (Sandy) Thomas, a former Senate energy aide and former president of H2Gen Innovations, a Virginia-based manufacturer of hydrogen production and purification equipment and now a respected consultant and analyst, summarized the issue in 2009 in a note to the author,[22] hydrogen has not taken off because society does not yet place value on sustainability:

In economic terms, the cost of fuels does not include the externalities of health effects due to urban air pollution, oil spills, ground water contamination, the military cost of defending oil, and, most important, the potential risks of major climate change. Put another way, society has a very high discount rate—we discount any adverse effects that occur in the future.

If the price of coal, oil, and, yes, even natural gas included a full accounting of externalities, then hydrogen would look much more promising overnight. If people had to pay $10/gallon for gasoline or 30 cents/[kilowatt-hour] for electricity to cover fossil fuel damages to our health and environment, then suddenly hydrogen fuel-cell vehicles and hydrogen produced by wind, solar or biomass would look like a bargain. Investors would flock to hydrogen equipment manufacturers. People would convert their SUVs to run on clean-burning hydrogen derived from wind energy at only $2.50/gallon of gasoline equivalent.

A truly sustainable energy future has two attributes: no pollution or greenhouse gas emissions, and no consumption of non-renewable resources. There are only two energy options that meet this sustainability goal: renewable hydrogen and fusion.

Pessimistically, Thomas added:

Sustainability requires the intervention of governments. Governments alone have the responsibility of protecting the commons. Industry has no major incentive (other than public relations) to build a sustainable energy system. Their overriding objective is return on investment, and burning fossil fuels is very profitable. At best, they will sponsor renewable energy R&D or fuel-cell programs with an infinitesimally small fraction of their profits to give the appearance of preparing for a sustainable future. But most governments do not have the vision or leadership to look into the future and to implement policies that will provide for the welfare of future generations.

Summarizing, Thomas said:

All the key decisions makers who could influence a transition to clean energy carriers like hydrogen have a very short time horizon: industries have to show a return on investment within a few years, and most elected officials feel that they must show results before the next election—at best six years for a Senator, four years for a President, and only two years for a Representative.

Reflecting on recent events, Thomas added:

We were all hopeful that President Obama would provide the necessary vision and leadership. Unfortunately he has been diverted by two wars, the worst economy since the Great Depression and a domestic agenda dominated by health care reform. As a result, he left the energy debate in the hands of Steven Chu, a Nobel laureate in physics who has the credentials and science background to understand the unique advantages of a transportation future built around hydrogen and fuel cells. Much to our shock and dismay, Dr. Chu zeroed out the hydrogen and fuel cell electric vehicle program as his first action as Secretary of Energy. As best we can determine, he made this short-sighted, devastating decision without consulting with any of the key players including automobile companies that have spent billions of their own dollars developing fuel cell electric vehicles over the last 15 years, the energy companies, or even his own Hydrogen Technical Advisory Committee (HTAC) that was set up to advise the Secretary on the hydrogen program. This action could set the hydrogen economy back another decade or two.

Earlier, he had asked:

Where do we find the visionary leaders who will look two or three decades into the future and imagine a better world for their children, grandchildren or even great grandchildren?

And the costs are changing. Fossil fuels will be harder to find and more expensive, and renewables are definitely getting cheaper. Warned Fatih Birol, the chief economist of the International Energy Agency, in an August 3, 2009, interview with Britain's *The Independent* newspaper, "One day, we will run out of oil, it is not today or tomorrow. . . . We will have to leave oil before it leaves us, and we will have to prepare ourselves for that day." The cost of storage and use technologies such as fuel cells are on a downward trajectory, though they have some way to go. Their advantages are forcing development in the right directions as the costs of conventional fuels are going up. Health and damage costs are much higher than ever before, and people are starting to consider them, though they may not be added to the price of a gallon of diesel.

Fears about global warming and CO_2 buildup in the atmosphere surfaced decades ago. In 1979, for example, a British Broadcasting Corporation TV documentary about hydrogen energy, "The Invisible Flame," quoted a meteorologist stationed in Hawaii, home of one of the world's most important atmospheric CO_2-monitoring posts, as follows: "We don't know at this point whether [CO_2] will build up so that it can do damage. The oil crisis may have slowed it a little. . . . A lot of people believe we could get into trouble, irreversible trouble, in about ten years' time."

Hydrogen contains no carbon at all. Burning it and converting it to energy produces no CO_2 and no greenhouse gas. Used as a fuel, it would reduce and eventually eliminate at least the man-made share of CO_2 deposited in the atmosphere. Switching to hydrogen energy—even using hydrogen from fossil fuels as a bridging measure—may help save our children's health and perhaps their lives; using hydrogen made from natural gas and used in fuel cell vehicles would reduce greenhouse gas emissions by approximately 50 percent right away.

The sky isn't falling—yet. But unless something is done on an international scale, with measures that prove we can actually use our collective human intelligence and wits to guarantee our survival, the time may come when the sky will turn so gray, poisonously yellow, or red from heat and pollution that it might as well be falling. Time will undoubtedly tell.

2

Hydrogen's Discovery: Phlogiston and Inflammable Air

Water is everything. So taught Thales of Miletos (a settlement on the western coast of Asia Minor). Thales, who lived from about 624 B.C. to 545 B.C., was a pre-Socratic Greek philosopher, reputedly the founder of the Milesian school of philosophy. Although he apparently never wrote anything, he was regarded as one of the Seven Wise Men of Greece in his time. The first Western philosopher of record, he is said to have introduced astronomy to ancient Greece. Before Thales, the universe was explained mostly in mythological terms. For Thales, however, water was the primordial material and the essence of everything else in the world. The ideas of Thales, said to be traceable to Babylonian beliefs, are "easily understandable in that the observation of water turning into rigid ice and its transformation into an air-like state led to the thought that all things were derived from matter of middle characteristics."[1] Other early philosophers added air (Anaximenos of Miletos), fire (Heracleitos of Ephesus), and earth (Empedocles of Agrigentum) to the list of elements.

In a way, Thales was not far off the mark. We know now that water consists of two elements: hydrogen and oxygen. Nevertheless, the preponderant part of water is hydrogen (in German, *Wasserstoff*—the stuff of water), the most abundant material in the universe and the simplest and lightest of the elements. It is believed to make up about 75 percent of the mass of the universe and to account for more than 90 percent of its molecules, according to the *New Columbia Encyclopedia*. Harvard astrophysicist Steven Weinberg says that 70 to 80 percent of the observable universe consists of hydrogen, and the rest is mostly helium.[2]

Hydrogen was first produced, more or less unwittingly, around the end of the fifteenth century, when early European experimenters dissolved metals in acids. However, its classification and description took about 200 years. Many scientists contributed to the unlocking of hydrogen's characteristics, an effort that was closely intertwined with the identification and chemical isolation of oxygen.

Not until the seventeenth century was doubt cast on the notion that air is one of the basic elements. A Dutch physician and naturalist, Herman Boerhaave (1668–1738), was the first to suspect some life-supporting ingredient in the air that is the key to breathing and combustion. "The chemists will find out what it actually is, how it functions, and what it does; it is still in the dark," he wrote in 1732. "Happy he who will discover it."[3] In England, the brilliant scientist Robert Boyle (1627–1691) also maintained that "some life-giving substance," probably related to those needed for maintaining a flame, was part of the air. The English physician and naturalist John Mayow (1645–1679) claimed that "nitro-aerial corpuscles" were responsible for combustion.[4]

The realization that both oxygen and hydrogen are gases was long delayed by the phlogiston theory, an early, erroneous attempt to explain the phenomenon of combustion. Promulgated by the German physician and scientist Georg Ernst Stahl (1660–1734) and first published in 1697, the theory held that a substance called phlogiston, which disappeared from any material during the combustion process, imparted burnability to matter. It was believed to be impossible to reduce phlogiston to a pure state. Modern chemistry tells us that to burn a material is to add a substance—oxygen—to it. Stahl held the reverse: that combustion was the release of phlogiston from the burning material. Similarly, he interpreted the reverse chemical reaction (reduction, in which oxygen is removed) as the addition of phlogiston. Even the increase in weight during oxidation, a fairly clear indication that something was added rather than removed, was explained in an altogether artificial fashion: Stahl claimed that phlogiston was so light that it was repelled by the Earth. When phlogiston was removed from a compound, Stahl claimed, the material gained weight because it had lost a component that had lightened it. Stahl, wrote one biographer, "did not hesitate to exclude facts if they violated his ideas: unity of thought was his ultimate goal above all factuality."[5]

Meanwhile, the British preacher Joseph Priestley (1733–1864), the Swedish-German apothecary Carl Wilhelm Scheele (1742–1786), and other scientists had discovered oxygen but had not named the element. Scheele isolated the burnable component of the atmosphere and labeled it "fire air." Sometime between 1771 and 1772, he was the first to produce pure oxygen. It was Scheele's bad fortune that his publisher put off publication of his major work, *Chemical Treatise on Air and Fire*, until 1777. His chief competitors, Priestley and Antoine Laurent Lavoisier, published their discoveries in 1774. In that year, Priestley discovered oxygen—he called it "dephlogisticated air"—when he heated mercury oxide without the presence of air. The resultant gas produced sparks and a bright flame in a glowing piece of wood kindling. When Priestley inhaled the gas, he "felt so light and well that he regarded it as curative and recommended it as a means of improving the quality of air in a room and as beneficial for lung diseases."

Priestley's and Scheele's experiments came to the attention of France's foremost chemist of the day, Lavoisier (1743–1794). Lavoisier, who had been studying gases for years, had noted that during burning, both phosphorus and sulfur absorbed part of the surrounding air and gained weight in that process. During a visit to Paris in October 1774, Priestley told Lavoisier about his experiments with mercury oxide. Lavoisier had recently received a letter from Scheele about his discovery of this gas, which makes flames burn "lively" and which "animals can breathe." Lavoisier repeated Priestley's experiments. In 1772, Lavoisier had been among the first to make precise weight measurements to quantify how much "air" disappeared during combustion of phosphorus and sulfur. In an elaborate twelve-day experiment, he had heated mercury and air in an airtight retort, producing that same gas that was so conducive to combustion and breathing. Lavoisier labeled this gas "oxygen." He concluded one of his papers as follows: "We shall call the change of phosphorus into an acid and in general the combination of any burnable body with oxygen as oxidation." In 1789, Lavoisier, not content to refute Stahl's phlogiston theory with experimental evidence, staged a play in Paris to destroy the theory completely. A German visitor wrote, "I saw the famous M. Lavoisier hold an almost formal auto-da-fé in the Arsenal in which his wife appeared as a high priestess, Stahl as advocatus diaboli to defend phlogiston, and in which poor phlogiston was burned in the end following the accusations by oxygen. Do not consider this a joking invention of mine; everything is true to the letter."[6]

The discovery of hydrogen as an element also proceeded by fits and starts. The Chinese reportedly doubted early on that water was an indivisible element. In the Middle Ages, the famous physician Theophrastus Paracelsus (1493–1541) was apparently the first to produce hydrogen when he dissolved iron in spirit of vitriol. "Air arises and breaks forth like a wind," he is reputed to have said of his discovery, but he failed to note that hydrogen was burnable. Turquet de Mayeme (1573–1655) noted hydrogen's burnability after he mixed sulfuric acid with iron—a phenomenon rediscovered by the French chemist and apothecary Nicolas Lemery (1645–1715), who described the burning of the gas as *"fulmination violente et éclatente."* Still, there was no thought that this gas was an element; rather, it was believed to be some sort of burnable sulfur.

The final isolation and identification of hydrogen was roughly concurrent with the unraveling of the secrets of oxygen in the second half of the eighteenth century, largely because the same scientists were investigating both air and water. Boyle, for instance, was researching artificial gases—"factitious air," he called them—and was producing hydrogen from diluted sulfuric acid and iron. Boyle did not regard these gases as significantly different from common air; he saw them as a type of air with different characteristics, a view shared by many chemists of his day.[7]

Henry Cavendish (1731–1810), an English nobleman, was the first to discover and describe some of hydrogen's qualities. However, Cavendish did not name the element hydrogen; caught up in the prevailing belief in phlogiston, he thought he had discovered phlogiston in a pure state—a belief he clung to until his death. Taking off from investigations of "factitious air" by other scientists, Cavendish found that there were two different types: "fixed air" (carbon dioxide) and "inflammable air" (hydrogen). Describing these findings in his first scientific paper, which he presented to the Royal Society of London in 1766, Cavendish gave precise readings of specific weight and specific density for both gases. He proved that hydrogen was the same material as "inflammable air," even though it was derived from different metals and different acids, and that it was exceedingly light—about one-fourteenth as heavy as air.

Hydrogen's buoyancy was quickly put to aeronautical use. "Our colleague has put this knowledge to practical advantage in making navigation in the air safe and easy," said a eulogizing contemporary the year after Cavendish's death.[8] He was referring to Jacques Alexandre César

Charles (1746–1823), a French physicist who confirmed Benjamin Franklin's electrical experiments and became interested in aeronautics. In 1783 Charles flew a hydrogen-filled balloon to an altitude of almost 2 miles. "In fact," said this contemporary "one can say that without Cavendish's discovery and Charles's application of it, the Montgolfiers' achievement would scarcely have been feasible, so dangerous and cumbersome for the aeronaut was the fire necessary for keeping ordinary air expanded in the montgolfières."[9]

Cavendish also demonstrated that mixing inflammable air (hydrogen) with air and igniting the mixture with an electric spark produced water and usually a remnant of air. In other experiments, he ignited hydrogen with pure oxygen; when the ratio was right, this yielded only water, thus establishing the makeup of that first "element." Cavendish's experiments involving electric sparks and hydrogen and oxygen, begun in the late 1770s, were not published until the mid-1780s, in his famous treatise *Experiments on Air.*

Lavoisier had been trying for some time to find out the nature of "inflammable air," which he also had obtained by dissolving metals in acid. On combustion of this gas, he expected to obtain an acid, but that was not the result. In 1783, Lavoisier heard of Cavendish's work through an intermediary (Charles Blagden, secretary of the Royal Society). Lavoisier immediately repeated the experiment, but his first attempt failed to impress fellow scientists with its significance. In other efforts, he took the reverse route: splitting water molecules in a heated copper tube. Iron filings in the tube turned black and brittle from the escaping oxygen, and "inflammable air"—a gas that could have come only from the water—emerged from the tube. In a landmark experiment, Lavoisier combined hydrogen and oxygen and produced 45 grams of water. (The water is still preserved in the French Academy of Science.) His classic, definitive experiments proving that hydrogen and oxygen constitute the basic elements of water were done before a large body of scientists in February 1785. In collaboration with other experimenters, he published his major work, *The Method of Chemical Nomenclature,* in which he labeled the "life-sustaining air" *oxygen* and the "inflammable air" *hydrogen.*

In 1793, four years after the storming of the Bastille, large-scale economical hydrogen production was invented under the shadow of the uprising and occasioned by the warfare of the competing factions, according to a historical account presented at the 1986 World Hydrogen Conference in Vienna.[10] Jean Pottier and C. Bailleux (of France's national utilities Gaz de France and Electricité de France, respectively) noted that Guyton de Norveau, a well-known chemist and "representative of the people" of the Comité de Salut Public (Committee for Public Salvation), suggested using hydrogen-filled captive balloons by the army as observation platforms. Norveau, together with Lavoisier, repeated Lavoisier's famous 1783 experiment on a larger scale, prompting the committee to approve the large-scale manufacture of hydrogen gas. The task was entrusted to another chemist-physicist, Jean Pierre Coutelle. Coutelle built a furnace equipped with a cast-iron tube, which he filled with some 50 kilograms of iron filings. Steam was piped in at one end, and hydrogen came out at the other—170 cubic meters of the gas in the first around-the-clock trial run, which lasted three days. Coutelle subsequently set up shop at an army camp at Meudon, close to Paris, where he built a forerunner of what today would be called a hydrogen generator.

The first action-ready generator was constructed in early 1794 at Maubeuge. Meanwhile, a collaborator named Conté refined the design into what Pottier and Bailleux called "the army's standard generator." Contemporary drawings mentioned by Pottier and Bailleux depicted a furnace with seven 3-meter-long iron tubes, each 30 centimeters in diameter, containing 200 kilograms of iron cuttings. Water was injected via a seven-way distributor, and the generated hydrogen was washed and cooled with a rotating washer behind which the inventors had installed a dryer-scrubber. The device also included a temperature-control system—seventy-five years before similar systems with similarly sophisticated components were devised for coal gas generators, according to Pottier and Bailleux.

In the early nineteenth century, so-called hydrogen gas was used to light and heat houses, hotels, and apartments and to supply street lighting. Usually this was not hydrogen at all but essentially carbon-containing gases derived from wood or coal. The confusion was due to the fact that all were lighter than air and were associated with the intrepid balloonists. (Pottier and Bailleux reported that in 1817 there was a Café du Gaz Hydrogène across from the Paris town hall, which in fact was lighted by coal gas.)

Lavoisier had been a member of the Ferme-Générale, a financial corporation that leased from the French government the right to levy certain taxes. The system was open to abuse, and some of its members were widely hated by the public. Lavoisier, who was also one of the commissioners in charge of gunpowder production for the government, got caught up in the swirl of charges and countercharges of the French Revolution, and he became one of its victims. In 1794 all members of the Ferme-Générale were convicted on trumped-up accusations, and Lavoisier went to the guillotine.

A History of Hydrogen Energy: The Reverend Cecil, Jules Verne, and the Redoubtable Mr. Erren

On November 27, 1820, the dons of Cambridge University assembled to hear a clergyman's proposal. It is recorded in the *Transactions of the Cambridge Philosophical Society* that Rev. W. Cecil, fellow of Magdalen College and the society, read a lengthy treatise, "On the Application of Hydrogen Gas to Produce Moving Power in Machinery," describing an engine operated by the "Pressure of the Atmosphere upon a Vacuum Caused by Explosions of Hydrogen Gas and Atmospheric Air." Cecil first dwelled on the disadvantages of water-driven engines (which could be used only "where water is abundant") and steam engines (which were slow in getting underway). The utility of a steam engine was "much diminished by the tedious and laborious preparation which is necessary to bring it into action." Furthermore, "a small steam engine not exceeding the power of one man cannot be brought into action in less than half an hour: and a four-horse steam engine cannot be used [without] two hours preparation." A hydrogen-powered engine would solve these problems, Cecil averred: "The engine in which hydrogen gas is employed to produce moving force was intended to unite two principal advantages of water and steam so as to be capable of acting in any place without the delay and labour of preparation." Rather prophetically, Cecil added, "It may be inferior, in some respects, to many engines at present employed; yet it will not be wholly useless, if, together with its own defects, it should be found to possess advantages also peculiar to itself."

According to Cecil's explanations, the general principle was that hydrogen, when mixed with air and ignited, would produce a large partial vacuum. The air rushing back into the vacuum after the explosion could be harnessed as a moving force "nearly in the same manner as in the common steam engine: the difference consists chiefly in the manner of forming the vacuum. . . . If two and a half measures by bulk of atmospheric air be mixed with one measure of hydrogen, and a flame be applied, the mixed gas will expand into a space rather greater than three times its original bulk."[1]

Cecil went on to discuss the workings of his engine in considerable detail. The *Transactions of the Cambridge Philosophical Society* did not record whether Cecil actually ever built such an engine. In any event, Cecil's proposal was the first known instance of an early technologist's attempting to put the special qualities of hydrogen to work.

Cecil's suggestion came only twenty years after another fundamental discovery: electrolysis (breaking water down into hydrogen and oxygen by passing an electrical current through it). That discovery had been made by two English scientists, William Nicholson and Sir Anthony Carlisle, six years after Lavoisier's execution and just a few weeks after the Italian physicist Alessandro Volta built his first electric cell.

In the next 150 years or so, the unique properties of hydrogen were discussed with increasing frequency by scientists and writers of early science-fiction. Probably the most famous example, well known in the world's hydrogen community, is Jules Verne's uncannily prescient description in one of his last books of how hydrogen would become the world's chief fuel. *The Mysterious Island* was written in 1874, just about 100 years before research into hydrogen began in earnest. In one remarkable passage, Verne describes the discussions of five Americans during the Civil War—Northerners marooned on a mysterious island 7,000 miles from their starting point of Richmond, Virginia, after a storm-tossed escape by balloon from a Confederate camp.[2] The five are the "learned, clear-headed and practical" engineer Cyrus Harding, his servant Neb, the "indomitable, intrepid" reporter Gideon Spillett, a sailor named Pencroft, and young Herbert Brown (an orphan and Pencroft's protégé). The five are discussing the future of the Union, and Spillett raises the specter of what would happen to commerce and industry if the coal supply were to run out:

"Without coal there would be no machinery, and without machinery there would be no railways, no steamers, no manufactories, nothing of that which is indispensable to modern civilization!"

"But what will they find?" asked Pencroft. "Can you guess, captain?"

"Nearly, my friend."

"And what will they burn instead of coal?"

"Water," replied Harding.

"Water!" cried Pencroft, "water as fuel for steamers and engines! Water to heat water!"

"Yes, but water decomposed into its primitive elements," replied Cyrus Harding, "and decomposed doubtless, by electricity, which will then have become a powerful and manageable force, for all great discoveries, by some inexplicable laws, appear to agree and become complete at the same time. Yes, my friends, I believe that water will one day be employed as fuel, that hydrogen and oxygen which constitute it, used singly or together, will furnish an inexhaustible source of heat and light, of an intensity of which coal is not capable. Some day the coalrooms of steamers and the tenders of locomotives will, instead of coal, be stored with these two condensed gases, which will burn in the furnaces with enormous calorific power. There is, therefore, nothing to fear. As long as the earth is inhabited it will supply the wants of its inhabitants, and there will be no want of either light or heat as long as the productions of the vegetable, mineral or animal kingdoms do not fail us. I believe, then, that when the deposits of coal are exhausted we shall heat and warm ourselves with water. Water will be the coal of the future."

"I should like to see that," observed the sailor.

"You were born too soon, Pencroft," returned Neb, who only took part in the discussion with these words.

Of course, Verne did not explain what the primary energy source would be to make the electricity needed to decompose water. But in the overall context of nineteenth-century scientific knowledge, Verne's foresight is remarkable.

Hydrogen also figures in a juvenile adventure novel that seems to have been published shortly after 1900 in England. A British scientist interested in hydrogen, W. Hastings Campbell, referred to the book briefly when introducing a hydrogen paper read in March 1933 at Britain's Institute of Fuel. Campbell told his distinguished audience that *The Iron Pirate* by Max Pemberton had made a great impression on him when he was a boy.[3] Pemberton's potboiler described the adventures of a gang of international crooks who owned a battleship that attained terrific speeds due to the use of hydrogen engines—"another instance of the very annoying persistence with which art always seemed to anticipate discoveries," said the account of that meeting in the *Journal of the Institute of Fuel*.

The 1920s and the 1930s witnessed a flowering of interest in hydrogen as fuel, especially in Germany and England but also in Canada. The evolution of Canada's Electrolyser Corporation—in its glory days, a leading maker of electrolytic hydrogen plants (it has delivered some 900 systems to ninety-one countries)—began early in the twentieth century. Around 1905, Alexander T. Stuart, the father of the last chairman, Alexander K. "Sandy" Stuart, began to take an interest in hydrogen energy while studying chemistry and mineralogy at the University of Toronto. Young Stuart and one of his professors, Lash Miller (a former student of the fuel cell's inventor, William Grove), had noted that most of Canada was importing almost all its fuel except for wood. "At the same time, Niagara Falls' hydroelectric generating capacity was being utilized at a capacity factor of only 30–40 percent," Sandy Stuart related in 1996 in the first of a series of lectures bearing his name at the University of Sherbrooke. "The question was, how could such surplus capacity be converted to fuel energy? The obvious answer was electrolysis of water. This led to our first experimental electrolysers."

As it turned out, Stuart electrolyzers came into commercial use not to make hydrogen fuel but to make hydrogen and oxygen for the purpose of cutting steel. The first electrolyzers were shipped in 1920 to what was then the Stuart Oxygen Company in San Francisco. Four years later, the Canadian government began supporting the use of Stuart electrolysis cells to make fertilizer in British Columbia. From the mid-1920s on, the elder Stuart also developed concepts for the Ontario Hydro utility to integrate hydroelectric energy with coal, coke, or other carbon sources to make "town gas" (carbon monoxide and hydrogen) for domestic heating, to produce a range of synthetic chemicals including methanol, and to directly reduce iron ore to iron. In 1934 Ontario Hydro built a 400-kilowatt electrolysis plant, and there were plans to heat buildings with hydrogen and even to run test vehicles on it, but that project was terminated after two years. All these efforts ended with changes in Ontario's governments, but mostly with Canada's entry into World War II and the arrival of natural gas on Canada's energy scene after the war.

On the conceptual level, one of the most important figures in those early years was John Burden Sanderson Haldane, a physiologist-turned-geneticist, longtime editorial board director of the communist newspaper the *Daily Worker*, and in the 1960s an émigré to India and a guru to that country's growing science establishment. In 1923, when he was in his late twenties, Haldane gave a famous lecture at Cambridge University in which he said that hydrogen—derived from wind power via electrolysis, liquefied, and stored—would be the fuel of the future.[4] In a paper read to the university's Heretics Society, Haldane said, "Liquid hydrogen is weight for weight the most efficient known method of storing energy, as it gives about three times as much heat per pound as petrol. On the other hand, it is very light, and bulk for bulk has only one-third the efficiency of petrol. This will not, however, detract from its use in aeroplanes where weight is more important than bulk." In the same paper, he prophesied that 400 years in the future, Britain's energy needs would be met by "rows of metallic windmills working electric motors which in their turn supply current at a very high voltage to great electric mains." "At suitable distances," he continued, "there will be great power stations where during windy weather the surplus power will be used for the electrolytic decomposition of water into oxygen and hydrogen. These gases will be liquefied and stored in vast vacuum jacketed reservoirs probably sunk into the ground. . . . In times of calm the gases will be recombined in explosion motors working dynamos which produce electrical energy once more, or more probably in oxidation cells."[5] "These huge reservoirs of liquefied gases," Haldane forecast, "will enable wind energy to be stored so that it can be expended for industry, transportation, heating, and lighting, as desired. The initial costs will be very considerable but the running expenses less than those of our present system. Among its more obvious advantages will be the fact that energy will be as cheap in one part of the country as another, so that industry will be greatly decentralized; and that no smoke or ash will be produced."

Also in Britain, Harry Ricardo, one of the pioneers in the development of the internal combustion engine, and A. F. Burstall were among the first to investigate the burn characteristics of hydrogen, and W. Hastings Campbell, the German Rudolf Erren (who spent most of the 1930s in England), and R. O. King (then with the British Air Ministry Laboratory) worked on hydrogen as a fuel.

In Germany, Franz Lawaczeck, Rudolf Erren, Kurt Weil, J. E. Noeggerath, Hermann Honnef, and other engineers and inventors were researching hydrogen and advocating its use as a fuel. Some of these men admitted to being influenced by Jules Verne. Lawaczeck, a turbine designer, became interested in hydrogen as early as 1907. By 1919 he was sketching concepts for hydrogen-powered cars, trains, and engines.

Some of his inspiration came from contact with his cousin J. E. Noeggerath, an American of German birth who worked in Schenectady, New York, and later in Berlin. Lawaczeck and Noeggerath collaborated in developing an efficient pressurized electrolyzer. In the 1930s, Lawaczeck was apparently the first to suggest that energy could be transported by hydrogen-carrying pipelines. Honnef dreamed of huge steel towers, up to 750 feet in height, each with as many as five 480-foot windmills producing up to 100 megawatts each, which would be stored in the form of hydrogen; however, his concepts were never developed beyond the construction of a 50-meter prototype tower.

In Italy, a 1937 article in the journal *Rivista Aeronautica* mentioned in passing the experimental efforts of the engineer A. Beldimano to adapt liquid hydrogen for use in aircraft engines.[6]

In the United States, Igor Sikorsky mentioned the potential of hydrogen as an aviation fuel in a 1938 lecture before the American Institution of Electrical Engineers in Schenectady.[7] After predicting the development of a new type of aircraft engine that would permit planes to fly at speeds of 500 to 600 miles per hour at altitudes of 30,000 to 50,000 feet, Sikorsky said, "If a method of safe and economical production and handling of liquid hydrogen were developed for use as a fuel, this would result in a great change, particularly with respect to long-range aircraft. This would make possible the circumnavigation of the earth along the equator in a non-stop flight without refueling. It would also enable an increase in the performance of nearly every type of aircraft."

One of the earliest and most fascinating efforts involving hydrogen was its use not only as a buoyancy medium but also a booster fuel for the Zeppelins, the huge German dirigibles that provided leisurely and elegant transatlantic air travel in the 1920s and the 1930s. Normally, these big sky ships carried large amounts of liquid fuel (usually a benzol-gasoline mixture) that was used to drive 12- or 16-cylinder engines, which typically propelled a Zeppelin at an altitude of 2,400 feet and a speed of not quite 75 miles per hour—if there was no headwind. Fuel economy was one problem for the Zeppelin; another was how to reduce buoyancy as fuel consumption reduced a ship's weight. According to a 1929 report by the Zeppelin Company, the rule of thumb was that a Zeppelin's captain had to blow off about 1 cubic meter of hydrogen for every kilogram of fuel burned up during a nonstop cruise, which typically lasted three to five days. Better fuel economy could be achieved by certain engine modifications, such as increasing the compression ratios, but the buoyancy problem persisted. The solution was as simple as it was ingenious: Why not burn the blow off hydrogen as extra fuel together with the main fuel supply? Zeppelin's engineers found that this was feasible. The addition of between 5 and 30 percent hydrogen to the main fuel at compression ratios as high as 10:1 produced substantially higher power output—as much as 325 brake horsepower (bhp), in comparison with the normal 269 bhp. It also achieved substantial energy savings. The test-bed findings were confirmed by an 82-hour, 6,000-mile cruise over the Mediterranean Sea in 1928, during which a fuel reduction of about 14 percent was achieved. Experimenting along the same lines with diesel engines, the Royal Airship Works in Great Britain found that it was possible to replace almost the whole of the fuel oil by hydrogen without loss of power. On a typical England-to-Egypt trip, an airship would have saved almost 5 tons of fuel oil, according to these experiments. However, neither the British nor the Germans appear to have applied these findings to routine flights to a significant extent.

One of the best-known hydrogen advocates of the 1930s and the 1940s was Rudolf Erren, a brilliant, visionary German engineer who had trucks, buses, submarines, and internal combustion engines of all kinds running on hydrogen and other fuels, conventional and unconventional. Erren engines were powering vehicles in sizable numbers in Germany and England. A flinty engineer from Upper Silesia (now part of Poland) with a pronounced disdain for academics and theoreticians, Erren formed his first company, Erren Motoren GmbH Spezialversuchsanstalt, in a grimy industrial section of northern Berlin in 1928. Two years earlier, he had begun to investigate hydrogen and its properties, pursuing an interest that went back to his childhood. When I visited him in Hannover in 1976, he told me that he, like W. Hastings Campbell, had read Pemberton's *Iron Pirate* as a child. As he recalled the book, it "described a pirate group that had kidnapped a German professor who had developed a hydrogen engine which made the pirates' ship much faster than other ships."

Erren had experimented with hydrogen while attending high school in Katowice. His interest in hydrogen carried over as a hobby through his university years in Berlin, Göttingen, and England. "During summer vacations when other students went on vacation," he recalled, "I worked in engine workshops to learn something because I wanted to know these things in practice. Theory alone doesn't work." In 1928 he won his first patents, one of them for a hydrogen engine. Erren presented his data at the 1930 World Power Conference in Berlin. He told me that the terms *Erren engine, Erren process,* and *Erren system,* now largely forgotten, were then officially recognized to differentiate his combustion process from any other.

In 1930, at the invitation of several British firms, Erren went to London to found the Erren Engineering Company. There he continued his work on developing advanced combustion processes that would permit hydrogen to be used alone as a fuel or as a clean-up additive to normal fuels. The technique of "Errenizing" any type of internal combustion process was apparently relatively well known in the 1930s, at least among automotive engineers. Essentially it meant injecting slightly pressurized hydrogen into air or oxygen inside the combustion chamber rather than feeding the air-fuel mixture via a carburetor into the engine, a method that commonly resulted in violent backfiring. Erren's patented system required special fuel injection and control mechanisms but left the other engine components intact. With hydrogen used as a booster, the Erren system eliminated backfiring and achieved much better combustion of hydrocarbons with higher output and lower specific fuel consumption.

Kurt Weil, a German-born engineer who was Erren's technical director in the 1930s and who in the 1970s was a professor emeritus at the Stevens Institute of Technology in New Jersey, said that the idea of not permitting hydrogen to come into contact with the oxygen of the air before entering the combustion chamber was representative of Erren's "genius." Weil, who had been in the forefront of promoting hydrogen in the 1970s, explained: "When the valves were closed we injected hydrogen, which had a supercharging effect." This engineering approach was revived in the early 1970s.

In the mid-1930s, Erren and Weil proposed to the Nazi government—which by then was concerned with economic self-sufficiency and with reducing Germany's dependence on imported liquid fuels—that most internal combustion engines be converted to the Erren multifuel system. In addition to using carbon-based fuels produced from Germany's plentiful coal by the Fischer-Tropsch and Bergius systems, it would be possible to use hydrogen produced with off-peak power from Germany's closely knit grid of electric power stations, which normally ran at only about 50 percent of capacity.[8] By 1938, when Weil had fled Germany and gone to the United States, about a hundred trucks were running between Berlin and the industrial Ruhr area in the west—a distance of some 350 miles— switching from one fuel to another along the way ("with the truck fully loaded, on a steep incline with a switch in the cockpit," Weil recalled). In an interview in the late 1970s, Weil recalled that the engine adaptation was not especially difficult, but it was easier for some engine types than for others: "For a six-in-line it was much easier than for a V-type engine." In regard to the conversion costs, he estimated that "under [late 1970s] conditions the cost would have been a few hundred dollars per engine." Erren believed that more than a thousand cars and trucks were converted to his multifuel system.

The German railway system tested a hydrogen-powered self-propelled rail car in suburban operations out of Dresden. The train was powered by a 75-horsepower six-cylinder gasoline engine. It was "much worn" and running harshly, according to a 1932 report by a Reichsbahn maintenance depot; however, when primed with hydrogen, the engine developed up to 83 bhp—an increase of 9.7 percent. Powered by pure hydrogen, the engine produced 77 bhp.

Errenization was catching on in England too. Erren converted Carter-Paterson delivery vans and buses with Beardmore diesel engines to run on hydrogen for better fuel consumption and less pollution. Erren told of an incident involving members of an Australian commission who spent two or three weeks in his shops checking his claims and his engines. Eventually the commission wanted to conduct an open-road speed test with a bus. The site chosen was a hill outside London. According to Erren:

The police there were always on the lookout because the gentlemen from a nearby club drove faster than the thirty miles per hour speed limit. Well, we wanted an official confirmation. . . . The police were pretty well hidden, but we saw them anyway, switched to hydrogen and instead of driving at 30 miles we did 50 or 52 miles up the hill. The police stopped us, told us that they had timed us with a stop watch and we had exceeded the speed limit, which we had to admit. We paid our fine but thanked them profusely, which in turn astonished them until we explained that we now had official proof of our claims.

In 1935, Erren made headlines in the popular British press with news that warmed the hearts of Jules Verne fans. "Secret Fuel to Smash Air Record" headlined the *Sunday Despatch* of March 24, 1935, subheading the one-column story, "Non-Stop Round the World with Liquid Hydrogen." The story reported that engines were being perfected "in secret" in London that would "enable aeroplanes to smash the distance record; make long flights in the stratosphere; and fly non-stop around the world without refuelling." The project never went beyond the concept stage, however. Four decades later, Erren recalled that the prototype plane, a Rolls-Royce–powered De Havilland, was "ready to go," but that the idea fell by the wayside because of disputes as to whether the attempt would be made from Britain or Germany.

Two other Erren inventions, the "oxy-hydrogen" submarine and the trackless torpedo, attracted some attention in Britain in 1942. The trackless torpedo, fueled by oxygen and hydrogen, was beguilingly simple in concept. Erren started from the realization that torpedoes leave tracks of exhaust gas bubbles. Because hydrogen and oxygen recombined into water vapor, condensing back into the seawater, no bubbles were formed, and thus there was no giveaway trail. And the oxygen-hydrogen-burning submarine eliminated almost entirely the need for big batteries and electric motors for underwater running. Instead, during diesel-powered surface runs, the sub's engine also drove an electrolyzer, generating oxygen and hydrogen, which were then stored under pressure. When diving and running submerged, the same diesel engine burned the oxygen and hydrogen without any exhaust bubbles. Weight savings from the elimination of batteries and electric motors translated into the ability to carry more fuel and extended the sub's range—by one report, to as much as 15,000 miles. The generated oxygen was a valuable reserve for the crew in an emergency, and the pressurized hydrogen could be used to blow out tanks for surfacing if other air supplies were exhausted.

Erren was repatriated to Germany in 1945 after the end of World War II. All his personal and business possessions in England had been confiscated during the war, and the papers of his company, Deutsche Erren Studien GmbH, in Berlin, had been lost in Allied bombing. After moving to Hannover, where he helped set up a trade association of German plastics manufacturers, he worked for several years as an independent consulting engineer specializing in pollution control, industrial combustion processes, and related areas. None of his engines seem to have survived the war years.[9]

During World War II, interest in hydrogen as a fuel picked up in some parts of the world where fuel supplies were threatened or cut off because of hostilities. In Australia, industrial use of hydrogen was considered early in the war because of wartime demands for fuel oil and because the oil fields in Borneo had been lost to the Japanese. Queensland's government became attracted to hydrogen after the coordinator for public works, J. F. Kemp, learned about hydrogen progress in England and Germany on a 1938 visit to Britain. After Kemp returned, he ordered some studies of his own. However, it was not until the last year of the war that another Australian engineer, J. S. Just, completed a report dealing with hydrogen production via off-peak electricity in Brisbane. The hydrogen was to be used mostly for trucks. The Queensland government authorized construction of an experimental high-pressure plant in Brisbane, but not much was heard about it. The Allied victory in 1945 and the return of cheap oil and gasoline brought hydrogen progress to a halt.[10]

Interest in Hydrogen Picks Up after World War II

Interest in hydrogen picked up again around 1950 in the context of fuel cells. Francis T. Bacon, a British scientist, developed the first practical hydrogen-air fuel cell (a development that was to be of great significance later in the American space program).

Also in the 1950s, a German physicist was becoming interested in hydrogen as an energy-storage medium. Eduard Justi, a distinguished German electrochemist at the University of Braunschweig, had been working for years on the development of more efficient fuel cells. In a 1962 monograph, *Cold Combustion—Fuel Cells*, Eduard Justi and a coworker, August Winsel, discussed the prospects of splitting water into hydrogen and oxygen, storing these gases separately and recombining them in fuel cells.[11] Justi later amplified his ideas in the 1965 book *Energieumwandlung in Festkörpern* (Vanderhoeck & Ruprecht), in which he proposed using solar energy to produce hydrogen along the Mediterranean and piping it to Germany and other countries.[12]

In 1962, John Bockris, an Australian electrochemist, proposed a plan to supply U.S. cities with solar-derived energy via hydrogen. Bockris, who in 1975 published *Energy: The Solar-Hydrogen Alternative* (Halstead), the first detailed overview of a future solar-hydrogen economy, says that the term *hydrogen economy*, which has multiple economic and environmental meanings, was coined in 1970 during a discussion at the General Motors Technical Center in Warren, Michigan.[13] Bockris, at the time a consultant to GM, was discussing prospects for other fuels to replace gasoline and thereby help to eliminate pollution, a subject that was beginning to creep into the public consciousness. Bockris related in his book that the group concluded that "hydrogen would be the fuel for all types of transports." GM did some early experimental work on hydrogen but apparently did not give it the attention—at least, not the degree of publicity—that Daimler-Benz gave it six years later.

In 1970 an Italian scientist, Cesare Marchetti, delivered a lecture at Cornell University in which he outlined the case for hydrogen in lay terms.[14] Marchetti, at the time head of the Materials Division of the European Atomic Energy Community's Research Center at Ispra in northern Italy and one of Europe's most persuasive hydrogen advocates, had been calling for the use of hydrogen in Europe and the United States since the late 1960s.[15] Hydrogen, produced from water and heat from a nuclear reactor, could free humanity from dependence on dwindling fossil fuels, Marchetti said to the audience: "The potential for hydrogen is very great, and a smell of revolution lingers in the air." Marchetti, who has the gift of putting complex relationships into simple terms, stated the hydrogen proposition as follows:

The reason why the studies of industrial utilization of nuclear energy have concentrated on the production of electricity is that a substantial 20 to 25 percent of the energetic input in a developed society is used in the form of electricity and that its production is lumped in large blocks where reactors can show their economies. But almost nothing has been done to penetrate the remaining three quarters of the energy input: food, fuel, ore processing and miscellaneous uses where society is geared to using a wide variety of chemicals.

The problem is to find a flexible intermediate, produced in large blocks in which nuclear heat can be stored as chemical energy and distributed through the usual channels. . . .

In my opinion, the best candidate to perform such a task is hydrogen: on one side hydrogen can be obtained from water, a cheap and plentiful raw material. On the other side, hydrogen can be used directly and very efficiently for:

1. ore reduction, as an alternative to coal,

2. home and industrial heat as an alternative to oil,

3. in chemical synthesis, in particular (for making) ammonia and methanol,

4. producing liquid fuels, such as methanol, for transport; ammonia and hydrogen themselves have potential in the future,

5. producing food, particularly proteins, via yeasts such as hydrogenomonas.

Points 1 to 4 cover most of the 80 percent of the energy input, excluding electricity. Point 5 can solve once and for all the problem of feeding a growing world population.

A few scientists and engineers who had come to the same general conclusion in their respective disciplines spread a similar message in lectures, papers, and articles in the United States. Derek Gregory, a British scientist working at the Chicago-based Institute of Gas Technology, had become interested in hydrogen as a clean substitute for natural gas and wrote a seminal article on the hydrogen economy for the January 1973 issue of *Scientific American*. He was strongly supported in his work by Henry Linden, founding president of the IGT and éminence grise behind many of the early hydrogen-related R&D efforts in the United States.[16] Bob Witcofski, a young researcher working for the National Aeronautics and Space Administration (NASA), had become aware of the exciting prospects of liquid hydrogen as a fuel for aircraft, including nonpolluting supersonic and hypersonic airplanes. Lawrence Jones, a particle physicist at the University of Michigan, had become interested in hydrogen both as an offshoot of his scientific work and because of the rising concern over automotive pollution. The late Larry Williams (he died in May 2008) , a cryogenic specialist at the Martin-Marietta Corporation, had recognized the usefulness of liquid hydrogen as a fuel. Bill Escher, a former rocket engineer, had come to appreciate the potential of hydrogen as a fuel through his involvement with the U.S. space program.

In the early 1970s, while General Motors, Ford, and Chrysler mostly ignored the potential of hydrogen as a nonpolluting car fuel (publicly, at least), it captured the attention and the enthusiasm of many American academics, engineers, and automotive enthusiasts. Beginning roughly with the work of the Perris Smogless Automobile Association in California, and with a hydrogen-powered car built by the University of California at Los Angeles that placed second in the 1972 Urban Vehicle Design Competition sponsored by General Motors and other companies, efforts to use hydrogen in cars and trucks sprang up in the United States, Germany, Japan, France, and even the Soviet Union.

The U.S. military was also looking into hydrogen as a fuel. An air force program begun in 1943 at Ohio State University eventually culminated in the use of combined liquid hydrogen and liquid oxygen as rocket fuel in the U.S. space program. In 1956, Lockheed began secret work on a long-distance high-altitude reconnaissance plane, a forerunner of the U-2. In a parallel program, the National Advisory Committee for Aeronautics, forerunner of NASA, was gathering engine flight data in a B-57 twin-jet bomber operating partially on liquid hydrogen. The navy had been investigating hydrogen as a fuel for a variety of ships and hydrogen plus oxygen as a fuel for a deep-diving rescue vessel that would be powered by fuel cells. One revolutionary idea tossed around in the mid-1950s was to use energy from nuclear reactors powering aircraft carriers to make liquid-hydrogen fuel for carrier-based airplanes.

One significant military effort of the 1960s was the Army's Nuclear-Powered Energy Depot, "an early experiment in the hydrogen economy," according to a paper presented in 1974 in Miami Beach at the first major international hydrogen conference, dubbed The Hydrogen Economy Miami Energy (THEME). The idea was to develop a portable nuclear reactor that could split water into hydrogen and oxygen in the field, making hydrogen available as a chemical fuel for battle tanks and trucks. It was an outgrowth of the "recognition that the dominant problem in the combat theater is the transportation of petroleum," said John O'Sullivan, then an army chemical engineer and in the 1990s the manager of a fuel cell program at the Electric Power Research Institute in Palo Alto. The idea was dropped because of efficiency problems and because a portable nuclear hydrogen plant was a "very vulnerable item" that "needed a lot of people" and lost its main advantage of mobility if it had to be buried or otherwise protected from attack.

The enthusiasm for hydrogen in the early 1970s was a by-product of growing environmental awareness (especially concern over automotive pollution and the mounting conviction that alternative nonpolluting transportation systems and energy forms were needed) and of the awareness that, with the main sources of petroleum thousands of miles away in the politically volatile Middle East, energy sources closer to home should be looked at.

The oil shock of 1973 announced that the age of cheap, convenient liquid fuel would be coming to an end at some point and that substitutes would have to be found. At first, hydrogen seemed to provide an easy, fairly fast answer. Produced by electrolysis "cheaply" from "safe, clean" nuclear reactors (so went the conventional wisdom then), hydrogen could be substituted readily for fossil fuels. Thus, environmental concern and the desire for energy security combined to speed up the investigation of hydrogen.

The reasons for the renewed interest in hydrogen were, of course, different for different people, but the idea of a "totally benign energy metabolism," as Lawrence Jones of the University of Michigan once put it, was certainly a large factor. Hydrogen, he observed, had "a kind of gut appeal to people."[17] Jones, a particle physicist, put it more formally in a 1971 article in *Science* magazine, writing that the possibility of using liquid hydrogen as an ultimate replacement for fossil fuels had occurred to him in a casual conversation "related to the logistics and use of large quantities of liquid hydrogen in a cosmic-ray experiment."[18] "In remarking on the drop in price of liquid hydrogen in recent years," he recalled, "I noted that the cost per liter was about the same as that of gasoline." As he began reading up on hydrogen in the available literature, he reported, "I recognized that . . . it had an inherent self-consistency and appeal which warranted broader discussion. The conclusion I have reached is that the use of liquid hydrogen as a fuel not only is feasible technically and economically, but also is desirable and may even be inevitable."

In another article, published two years later in the *Journal of Environmental Planning and Pollution Control,* Jones wrote, "It soon became apparent that a surprising number of widely separated individuals and groups had very similar thoughts."[19] That phenomenon broke into the open in 1972 at a spring meeting of the American Chemical Society in Boston where Cesare Marchetti and a coworker, Gianfranco De Beni, presented their first thermochemical water-splitting process, and again at the Seventh Intersociety Energy Conversion Engineering Conference in San Diego that autumn.

In its September 23, 1972, issue, *Business Week* ran a multipage article on international hydrogen developments.[20] (Its effect on the scientists assembled in San Diego was apparently riveting. Marchetti later wrote, in a personal note, that "out of 650 participants about 500 were concentrated in the [session] on H_2.") *Fortune* carried a longer story two months later, and that was followed by articles in *Readers Digest, Time, Popular Science,* and other periodicals.

Hydrogen researchers' recognition that they were not alone reached a climax of sorts on May 6, 1972, the thirty-fifth anniversary of the May 6, 1937, Hindenburg airship disaster when the hydrogen-buoyed lighter-than-air dirigible caught fire, burned, and crashed during its landing attempt at Lakehurst Naval Air Station near Lakehurst, New Jersey (see chapter 11) with the creation of the informal H_2indenburg Society, a group dedicated "to the safe utilization of hydrogen as a fuel." Bill Escher, whose name appeared on many of the papers discussing hydrogen published in the United States in the 1970s, was the society's secretary.[21]

At the 1974 THEME conference, the groundwork was laid for setting up the International Association for Hydrogen Energy, which has been sponsoring biannual World Hydrogen Energy Conferences ever since. T. Nejat Veziroglu, president of the International Association for Hydrogen Energy, recounted the following in 1994 in his opening remarks at the conference in Cocoa Beach: "In the afternoon of the second day a small group, later to be named 'Hydrogen Romantics,' got together: Cesare Marchetti, John Bockris, Tokio Ohta, Bill Van Vorst, Anibal Martinez, Walter Seifritz, Hussein Abdel-Aal, Bill Escher, the late Kurt Weil, myself and a few other enthusiasts."[22] After a "passionate, yet deliberate debate," it was agreed "that the Hydrogen Energy System was an idea whose time had arrived." "It was a permanent solution to the depletion of conventional fuels, it was the permanent solution to the global environmental problem," Veziroglu said. "It was Anibal Martinez of Venezuela—incidentally, one who took part in setting up the petroleum cartel OPEC—who urged the founding of a society dedicated to crusade for the establishment of the inevitable and the universal energy system," Veziroglu added. "The rest is history." Officially chartered in the autumn of 1974, the association had about 2,500 members by 2009. In 1976 it began publishing a quarterly, which soon turned into a bimonthly and is now a monthly peer-reviewed scholarly journal, the *International Journal of Hydrogen Energy.*

Governments and international organizations were beginning to take notice. In the United States, where hydrogen research funding did not pass the million-dollar-per-year mark until the mid-1970s, $24 million was budgeted for hydrogen research in fiscal 1978—far too little, in the opinion of hydrogen advocates, who compared it to the $200 million the recently created Department of Energy (DoE) had allocated that same year for research on how to convert coal into natural gas.[23] Both the disparity and the sentiment have changed little: after declining to slightly more than $1 million in the early 1990s, the DoE's hydrogen program budget had slowly climbed back up to $24 million by 1999.

West Germany also began funding hydrogen programs on a small scale, earmarking $2 million between 1978 and 1980. Beginning in 1976, the Paris-based International Energy Agency began to support hydrogen programs; in 1978, its hydrogen budget stood at somewhat more than $16 million, spread over several years.

Starting in 1972, the European Economic Community spent an esti-mated 60 to 70 million units of account (roughly $72 million to $84 million) on hydrogen. But with interest in hydrogen cooling again as the oil shock began to fade later in the decade, the European Community cut back, budgeting 13.2 million units of account (about $10.2 million) for the period 1975 to 1979.

Japan's Project Sunshine, a twenty-six-year undertaking begun in 1974, initially seemed to be a truly gigantic alternative energy program, comparable in scope to the Apollo Program. It was supposed to assure Japan of clean and plentiful power by the year 2000. Early reports cir-culating among Western European scientists in the mid-1970s mentioned total outlays of up to $15 billion over the lifetime of the program, includ-ing a hydrogen budget of $3.6 billion.[24]

Shifting Attitudes: Growth, Some Disillusionment

A perceptible shift in attitude took place among the hydrogen workers as their ranks swelled in the 1970s. At the end of the 1960s and in the early 1970s, a relatively small number of highly idealistic individuals, scattered on different continents, formed a kind of elite international movement. They worked more or less independently, often as an after-hours labor of love. Communication among the members was spotty. Beginning perhaps in the mid-1970s, the scene began to change: more information about hydrogen began to percolate through the international scientific community, corporations, and energy planners as more and more researchers and institutions took up the cause full time. Because of that growth, institutionalization set in, and with it came some disillusionment. In part, the disillusionment had to do with excessively high initial expectations. American researchers and enthusiasts in particular grabbed onto the idea of a hydrogen economy as an almost instant panacea for both energy security problems and environmental problems. When neither a real hydrogen economy nor the beginnings of hydrogen hardware evolved rapidly within a few years, a letdown was probably inevitable, especially in view of the American penchant for putting new ideas into action immediately. The Europeans were perhaps more realistic (or slower on the uptake, depending on one's point of view). Marchetti and others had argued all along that any fundamentally new energy system required decades—maybe a century—before it would take over a sizable share of the total energy pie.

At the 1976 Miami Beach hydrogen conference, keynote speaker Derek Gregory said that he was more pessimistic about the chances for quick introduction of hydrogen as a universal fuel than he had been two years earlier at THEME. NASA's Bob Witcofski, who had also been at THEME, agreed but remained enthusiastic: "Some day, I don't know when, the hydrogen economy is going to come. And I would like, years from now, whenever it is, to be able to stand up and say, all right, our hydrogen airplane has improved 20 percent over the first crude model that was demonstrated back in the year 1990 by a Polack who worked for NASA—and I would like to be that Polack."

As it turned out, neither Witcofski nor anybody else at NASA ever got around to building that "first crude model" during the 1990s. However, the Russians did, in the late 1980s—and so did an octogenarian aviation pioneer in Florida. The Russians did it by adapting one of three engines of a commercial jetliner to liquid hydrogen. William Conrad, a retired Federal Aviation Administration air transport rating examiner (and Pan American's first director of flight training), did it by converting a small single-engine plane to liquid hydrogen.

In general, the 1980s, at least the first half, were not good for hydrogen—or for alternative, renewable-energy, or environmental issues in general. With the advent of the Reagan administration, the environment and clean energy were relegated to a back burner. The order of the day was beefing up the military (cost be damned), so as to confront the Soviet Union and fight communism around the globe, and promoting unrestrained growth for the private sector (including fossil fuel industries) at the expense of larger and ultimately much more important societal issues. Charts based on Congressional Budget Office numbers and prepared in the office of Senator Tom Harkin (D, Iowa), showed that the Reagan administration cut renewable-energy budgets by 80 percent almost immediately. At the same time, spending on nuclear weapons went, in constant 1992 dollars, from about $5 billion to about $10.5 billion in 1990, peaking at around $12 billion in 1992 (presumably because of budget commitments dating from the Reagan years).

Still, some people never gave up hope and kept up the battle for clean-energy technologies. It is difficult to say exactly when the tide began to turn again, but one landmark event came in spring 1986 with the first congressional hearings on hydrogen as fuel in ten years. The hearings were called by Representative George Brown Jr. (D, California), chairman of the House Science Committee's Subcommittee on Transportation, Aviation, and Materials, to consider the Hydrogen Research and Development Act, a bill he had introduced earlier; he was at the time the principal and perhaps the only hydrogen supporter in the House.[25] Brown launched the bill in part to capitalize on the attention given to the hydrogen-fueled aerospace plane—the ultimately ill-fated "Orient Express"—championed by, of all people, President Ronald Reagan in his State of the Union address earlier in 1986. Brown's bill would have authorized a $200 million, five-year DoE hydrogen research program, established a technical panel on hydrogen, required better interagency coordination on hydrogen research, and required NASA to set up a comprehensive hydrogen research program. Said Brown, "The major problems in using hydrogen appear to be economic, primarily related to production. New ideas . . . point to potential breakthroughs that may dramatically change the economics of hydrogen production." Predictably, the Reagan administration's witnesses were not impressed. Robert San Martin, the DoE's deputy assistant secretary for renewable energy, dismissed the idea, saying "new legislation is not required."[26]

The low point in official interest in hydrogen on part of the U.S. government came in 1987, when the DoE proposed a measly $1 million for hydrogen research and development for fiscal year 1988.[27]

Brown and Senator Spark Matsunaga (D, Hawaii) tried again in May 1987, simultaneously introducing new five-year, $200 million hydrogen bills and companion bills targeting faster commercialization of fuel cells. Congress did not pass the legislation until 1990 (the landmark Spark M. Matsunaga Hydrogen Research, Development and Demonstration Act, named in honor of the senator after his death).

During the next decade, progress seemed painfully slow to observers of the international hydrogen scene, resembling the kind of forward lurches often ascribed to the dialectical materialism school of history— two steps forward followed by one backward. Still, it was obvious that a lot had in fact happened during that decade and that progress had been made. For example, Daimler-Benz unveiled its first consumer-friendly fuel cell minivan in Berlin in the summer of 1996, followed in the fall by Toyota's RAV4 fuel cell vehicle shown at an electric vehicle symposium in Osaka. In the United States, Dan Sperling, director of the Institute for Transportation Studies at the University of California, Davis, told a House science subcommittee that the automotive Big Three, members of the Partnership for a New Generation of Vehicles (PNGV), should be required to more vigorously pursue low-polluting technologies such as fuel cells.

In 1997, Ford announced plans for building three hydrogen proton exchange membrane fuel cell prototype cars in cooperation with Ballard Power Systems and other partners in a partnership that also involved Daimler-Benz. And in one of too many instances of hyped but ultimately deflated expectations, a Northeastern University husband-and-wife team, chemists Nelly Rodriguez and Terry Baker, claimed in early 1997 they had developed an onboard hydrogen storage technology that permitted driving ranges of some 5,000 miles for fuel cell cars. Their claims could not be duplicated and quickly evaporated. Daimler-Benz Aerospace AG, the carmaker's aerospace division, said it would start work on a liquid hydrogen–fueled transport airplane. In Washington, the top-level President's Council of Advisors on Science and Technology (PCAST) said in November 1997 that the government's hydrogen program should be roughly doubled (The committee's energy panel that recommended this was chaired by Harvard environmental policy professor John P. Holdren, who in 2009 became President Obama's adviser for science and technology and director of the White House Office of Science and Technology Policy, as well as cochair of PCAST). Clinton himself called for a $6 billion global warming R&D program in his 1998 State of the Union speech "to encourage innovation, renewable energy, fuel efficient cars, energy-efficient homes," moves endorsed that spring by a Union of Concerned Scientists report urging Congress to support alternative fuel technologies, notably hydrogen-powered fuel cell buses. And while the United States and other countries kept on talking and writing about hydrogen and fuel cell energy, a little country with a population of some 300,000 on Europe's periphery, Iceland, decided to do something about it. Beginning that year, Iceland planners, in cooperation with Daimler-Benz and Ballard, began laying the groundwork for what eventually was supposed to be a national hydrogen economy.

In spring 1999, Iceland, with assists from what by then had morphed into DaimlerChrysler; Shell, and Norsk Hydro, formed a new company, the Icelandic Hydrogen and Fuel Cell Company, to move ahead with hydrogen technology with the ultimate goal of running the country's cars and buses and, most important, this seafaring nation's fishing fleet on hydrogen and fuel cells. By 2008, these ambitions had progressed to the point that Iceland's Hertz rental agency began to advertise the availability of hydrogen rental cars—Toyota Prius hybrids converted to hydrogen—and a tourism excursion boat was equipped with a silent auxiliary fuel cell. But these ambitions slowed with the global economic meltdown, which hit Iceland harder than most other countries. The goal is still to have as many as 600 hydrogen cars running by 2015, but with deliveries of more vehicles slowing, it was unclear whether this particular target could be met. However, as part of a widened Scandinavian effort, Iceland—along with Denmark, Norway, and Sweden—was cosigner of an early-2011 Memorandum of Understanding with Korean carmaker Hyundai-Kia to deploy hydrogen fuel cell vehicles in the region starting around 2015.

Encouraging Signs at the Turn of the Century

As the twentieth century drew to a close, the signs were encouraging. Europe's first hydrogen gas stations opened in Hamburg and Munich in 1999. Time magazine named two pioneers, Ballard Power Systems founder, the late Geoffrey Ballard, and maverick self-taught inventor/scientist/hydride storage specialist and fuel cell designer Stanford Ovshinsky as the magazine's first "Heroes of the Planet." California governor Gray Davis launched the California Fuel Cell Partnership with Ford and DaimlerChrysler; fuel cell maker Ballard Power Systems; oil companies ARCO, Shell, and Texaco, and two state agencies. Ford showed off its new hydrogen refueling station in its Dearborn r&d complex, unwrapped an experimental hydrogen-fueled internal combustion engine as possible bridge to a fuel cell future, and let reporters take a crack driving its first fuel cell car, the P2000.

The start of the new century saw German appliance maker Vaillant announce plans for a large European field test of residential fuel cells, and Ford showed off a prototype fuel cell version of its new Focus compact sedan. Scientists from the National Renewable Energy Laboratory and the University of California, Berkeley, developed an experimental method of renewable hydrogen production from green algae, *Chlamydomonas reinhardtii,* and the DoE issued a five-year timetable in its Hydrogen Infrastructure Blueprint. In an effort labeled "You Go H2 Yugo!!" California enthusiast Winstead Weaver showed off his 1986 Yugo ("a museum piece," he said) reengineered with a hydrogen-powered surplus 25-horsepower gas turbine which generated the electricity to run the 20-horsepower electric traction motor, to bemused DoE officials at the annual DoE Hydrogen Review meeting in San Ramon, California. In Alaska, the U.S. Postal Service dedicated the biggest-yet fuel cell plant, a 1-megawatt unit consisting of five 200-kilowatt phosphoric acid fuel cells made by International Fuel Cells to power the facility next to Ted Stevens Anchorage International Airport. In Las Vegas, the world's first small hydrogen fuel cell locomotive, a 14-kilowatt underground mining locomotive designed by the Denver-based Fuel Cell Propulsion Institute, was unveiled at an international mining technology trade fair. NASA announced plans to test a hydrogen-fueled Mach 7 hypersonic scramjet engine.

The following year, 2001, saw New England's governors and five eastern Canada premiers sign off on a climate action plan for the North American Northeast to reduce greenhouse gas emissions, a move that many environmentalists saw as a slap on the wrist of the do-little Bush administration. *Time* magazine included a sleek Italian prototype fuel cell bike designed by motorcycle builder Aprilia as one of the "Inventions of the Year" (it was supposed to become commercially available in 2003 but apparently never did). Honda installed a solar-powered hydrogen fueling station at its facilities in Torrance, California, the first for any carmaker, and Boeing announced plans for an experimental small fuel cell–powered plane.

The next year, General Motors rolled out its radical fuel cell AUTOnomy car concept at the Detroit Auto Show, the first of a series of concept vehicles that culminated five years later in the launch of a fleet test, Project Driveway, in which some 100 fuel cell Equinox sport utility vehicles were turned over to everyday drivers to gauge their reaction and collect operating data.

Timeline: International Activities since the Late 1980s

It would take too long to describe all the convoluted steps of international hydrogen activities since the late 1980s, but here are a few more highlights, some of which will be discussed in more detail in later chapters:

1987

• A two-year study commissioned by Canada's parliament and by two federal ministries urged Canada to make hydrogen energy technology a "national mission."

1988

• The Soviet Union's Tupolev Design Bureau converted a 164-passenger TU-154 commercial jet partially to liquid hydrogen, operating one of the three engines on the rocket fuel. The 21-minute maiden flight took place April 15.

• In May, in Fort Lauderdale, William Conrad became the first person to fly an airplane (a four-seat Grumman American Cheetah) exclusively fueled by liquid hydrogen.

- The previously unknown high level of hydrogen research activities in the former Soviet Union was highlighted by the first World Hydrogen Energy Conference held in Moscow. Almost half of the conference's 150 or so papers came from Soviet scientists. One seminal paper by two American scientists, Joan Ogden and Robert Williams of Princeton University's Center for Energy and Environmental Studies, predicted that hydrogen produced by means of photovoltaic cells could be economically competitive with coal-based synthetic fuels and even with electricity early in the twenty-first century.
- Ontario's Energy Ministry launched a $600,000 (Canadian dollars) program for the first tests of a novel fuel cell, a solid polymer electrolyte fuel cell developed by Ballard Technologies and Dow Canada.
- In Germany, sea trials got underway of a submarine powered by a hydrogen-fueled alkaline fuel cell developed by Siemens.
- The European Community and Canada's Province of Quebec agreed in December on the Euro-Quebec Hydro Hydrogen Pilot Project, a joint study of the feasibility of shipping electrolytically generated hydrogen from Quebec to Europe.

1989

- The National Hydrogen Association was launched in Washington with about a dozen members. By the late 1990s, its membership had grown to more than sixty.
- After trying for more than ten years, two scientists at the Geophysical Laboratory of the Carnegie Institution in Washington announced they had produced hydrogen in a metallic state. Ho-Kwang Mao and Russell Hemsley said they had achieved the feat by compressing the gas under ambient conditions to pressures of more than 2.5 megabars (2.5 million atmospheres).
- An international committee to establish technical standards for hydrogen energy was established in Zurich with the blessings of the International Organization for Standardization.

1990

- The world's first experimental solar-powered hydrogen-production plant became operational at Solar-Wasserstoff-Bayern, a research and testing facility in southern Germany.

- West Germany and the Soviet Union agreed to jointly develop propulsion technology for a liquid-hydrogen-powered prototype jetliner.
- A study by the University of Miami's Clean Energy Research Institute estimated that the environmental costs of burning fossil fuels amounted to about $2.3 trillion for 1990—equivalent to about $460 for every man, woman, and child on the planet. For the United States, the researchers said this was equivalent to adding roughly $1 to each gallon of gasoline in real costs.[28]
- ACHEMA, the world's premier triannual chemical equipment industry show in Frankfurt, added a hydrogen energy section to its exhibits.
- General Motors's Allison Gas Turbine Division began work on a methanol-fueled 10 kilowatt proton exchange membrane (PEM) fuel cell in conjunction with the Los Alamos National Laboratory, the Dow Chemical Company, and Ballard Power Systems.

1991

- The first test runs of an experimental liquid-hydrogen refueling device for cars and buses began at Solar-Wasserstoff-Bayern's prototype solar hydrogen plant in Bavaria. The goal was to refuel a car in a few minutes.

1992

- A grid-independent solar house that used hydrogen for long-term energy storage in Freiburg, Germany, became operational. The house was designed by the Fraunhofer Institute for Solar Energy Systems.

1993

- Ballard Power Systems rolled out the world's first PEM fuel cell bus at its facilities in Vancouver.
- Japan unveiled a plan that called for spending about $2 billion over almost three decades to promote hydrogen-based clean energy internationally via its so-called WE-NET (World Energy Network) Project.
- Daimler-Benz and Canada's Ballard Power Systems began a cooperative effort to develop fuel cells for cars and buses.

- The South Coast Air Quality Management District, the air-pollution watchdog agency for the Los Angeles Basin, funded a study to look at fuel cell locomotives as a means of reducing pollution.

1994

- Clean Air Now, a nonprofit organization in Los Angeles, won $1.2 million in matching federal funds to set up the first photovoltaics-to-hydrogen demonstration project in North America to produce gaseous hydrogen from solar energy and water to fuel three pickup trucks.
- Daimler-Benz displayed its first NECAR I experimental fuel cell vehicle at a press conference in Ulm, Germany, and New Jersey's H Power Corporation launched its methanol-fueled phosphoric acid fuel cell bus during Earth Day activities in Washington, D.C.
- The Saudi-German HYSOLAR demonstration plant, near Riyadh, went into operation.
- The first of four hydrogen-powered buses developed under the Euro-Quebec Hydro-Hydrogen Pilot Project was unveiled in Geel, Belgium.

1995

- Five vehicles running on hydrogen, hydrogen blended with other fuels, or a fuel from which hydrogen was extracted were displayed at the Sixth Annual U.S. Hydrogen Meeting in Alexandria, Virginia, organized by the National Hydrogen Association.
- The Chicago Transit Authority announced plans for long-term testing of three hydrogen-fueled Ballard PEM fuel cell buses.
- Daimler-Benz Aerospace announced plans for an experiment in which one of the two engines of a DO-328 commuter airplane would be converted to run on liquid hydrogen.

1996

- A stationary molten-carbonate fuel cell power plant operating on natural gas and designed by the Energy Research Corporation began generating electricity in Santa Clara, California.
- At an international press conference in Berlin, Daimler-Benz unveiled the prototype of the first consumer-friendly, pollution-free fuel cell passenger car: the NECAR II minivan.
- Toyota unveiled an experimental PEM fuel cell version of its popular RAV4 sport utility vehicle.

- *Element One,* an hour-long American documentary on hydrogen energy, premiered at the World Hydrogen Energy Conference in Stuttgart.
- A Twentieth Century Fox movie about clean abundant hydrogen, *Chain Reaction,* much anticipated in the international hydrogen community, opened to general disappointment.

1997

- At the Detroit Auto Show, the Chrysler Corporation unveiled the mockup of a PEM fuel cell passenger car that would be fueled by hydrogen extracted from gasoline or other liquid hydrocarbon fuel by an onboard processor developed by Arthur D. Little.
- Addison Bain, a widely respected hydrogen supporter and a retired NASA engineer, challenged the belief that hydrogen was the cause of the Hindenburg disaster in 1937. Bain said that although hydrogen had obviously contributed to the blaze, the disaster was almost certainly caused by a spark and significant electrostatic activity in the atmosphere at the time, which ignited the impregnated, highly flammable skin of the dirigible.
- Daimler-Benz and Ballard Power Systems announced plans to spend more than $300 million for the joint development and eventual production and marketing of fuel cells for transportation.
- Daimler-Benz unveiled a third-generation PEM fuel cell vehicle: the experimental NECAR III, powered by a 50-kilowatt Ballard fuel cell running on methanol.
- Ford Motor Co. joined the Ballard/Daimler-Benz team for the development of commercial fuel cells for cars, trucks, and buses.
- The President's Committee of Advisors on Science and Technology, suggesting that hydrogen might become an energy carrier "of importance comparable to electricity" in the twenty-first century, recommended that the U.S. DOE's hydrogen program be expanded substantially.

1998

- General Motors and its German subsidiary, Opel, launched a new global PEM fuel cell project for automobiles in Europe. An experimental fuel cell version of Opel's new Zafira minivan was unveiled at the Paris Auto Show.

- Norsk Hydro, Norway's premier utility, chemicals, and metals conglomerate, announced plans to produce electricity by using hydrogen gas as a fuel for steam turbines.
- The Royal Dutch/Shell Group set up the new International Renewables division and a hydrogen team to investigate business opportunities in this area.

1999

- Royal Dutch/Shell set up a hydrogen division.
- DaimlerChrysler showed a fuel cell–powered Jeep concept vehicle. Ford unveiled a fuel cell version of its P2000 research vehicle and a mock-up of a fuel cell–equipped SUV.

2000

- Ballard Power Systems unveiled the world's first production-ready PEM fuel cell for automotive use at the Detroit Auto Show. The 70-kilowatt unit was to be the basic design to be used by carmakers such as DaimlerChrysler and Ford in the first generation of fuel cell cars, promised for 2004, and General Motors showed a fuel cell demonstrator version of a new hybrid five-passenger car, the Precept.
- At the Frankfurt Auto Show, DaimlerChrysler offered to be the first to deliver commercial fuel cell buses to transit agencies, with the initial deliveries scheduled for 2002.
- Korean carmaker Hyundai signed agreement with International Fuel Cells to develop demonstrator hydrogen fuel cell SUVs.
- Manhattan Scientifics unveiled a fuel cell bicycle, targeting markets in developing countries.
- NASA launched a hydrogen airplanes study, including possible use of fuel cells for propfan engines.
- Honda unveiled its experimental hydrogen fuel cell FXC-V3 car in Japan in November, saying it would show the car the following month in California at the opening of the new facilities for the California Fuel Cell Partnership in Sacramento.
- Trade organization Fuel Cells Canada was launched.

2001

- The United Nations Global Environment Facility gave the go-ahead for fuel cell bus demonstration projects in five developing countries.

- BMW started a world tour of ten liquid hydrogen-fueled sedans in oil-rich Dubai.
- Toyota and Mazda introduced new fuel cell vehicles for road testing in Japan.
- Nine European cities signed contracts with DaimlerChrysler for up to thirty hydrogen fuel cell transit buses in demonstration projects.
- International Fuel Cells sold the first fuel cell power plant in South America and announced a record-breaking 1.2-megawatt fuel cell power plant for Connecticut.
- GM/Opel set fuel cell car endurance records at GM's testing grounds in Arizona.
- Honda installed a solar-powered hydrogen fueling station at its R&D center in California, the first by a any carmaker.
- Three major Scandinavian companies announced joint plans for an experimental wind-hydrogen plant on Norway's west coast.
- Ballard Power Systems announced its first commercial PEM fuel cell system and its biggest sale of fuel cells and components to a carmaker.
- Boeing announced plans for a small experimental fuel cell plane.

2002
- General Motors rolled out its radical fuel cell AUTOnomy concept car at the Detroit Auto Show.
- The science adviser to British prime minister Tony Blair called for a ban on fossil-fueled cars and urged a massive green R&D research effort.
- A German shipyard christened world's first fuel cell–powered submarine.
- UTC Fuel Cells sold seven 200-kilowatt fuel cell power plants for the biggest fuel cell facility yet.
- Honda and Toyota were in a race to put the first fuel cell cars on American and Japanese roads.
- A prototype fuel cell mining locomotive performed well in a real-world test.
- General Motors unveiled its latest-generation Hy-wire fuel cell car at the Paris Auto Show.

- Hydrogen fueling stations opened in Richmond, California; Las Vegas, Nevada; and Berlin, Germany.
- DaimlerChrysler announced plans for a sixty-vehicle hybrid fuel cell car fleet for the United States, Europe, Japan, and Singapore.
- The European Commission convened the first meeting of a new high-level hydrogen/fuel cell working group, including Nobel Prize winner Carlo Rubbia of Italy.
- Toyota and Honda launched California's first fuel cell fleets in back-to-back events.
- Toyota's Hino subsidiary unveiled Japan's first fuel cell bus at the Tokyo Motor Show.
- U.S. Energy Secretary Spencer Abraham unveiled a National Hydrogen Energy Roadmap.

2003

- President Bush proposed $1.2 billion for hydrogen research in his State of the Union message.
- Japan's first hydrogen/fuel cell test facility opened.
- General Motors and Shell Hydrogen teamed up in a two-year, six-vehicle fuel cell demonstration project in Washington, D.C.
- The world's first commercial hydrogen fueling station opened in Reykjavik, Iceland.
- U.S. Energy Secretary Abraham proposed launching the International Hydrogen Energy Partnership.
- Dow teamed with General Motors in a 500-unit, 35-megawatt industrial fuel cell park, the world's largest.
- DaimlerChrysler turned over the first fuel cell city bus to Madrid, with twenty-nine more to come for other European cities.
- The U.S. DoE announced its "Grand Challenge" to develop better hydrogen storage technologies.
- Fifteen countries and the European Union signed a historic hydrogen cooperation pact.

2004

- California's plans for an interstate hydrogen fueling station network began to take shape.

- New Jersey developers started work on converting the first U.S. home to solar hydrogen, including seasonal storage.
- California's South Coast Air Quality Management District planned to convert thirty-five hybrid Prius cars to hydrogen.
- An international consortium started a five-year 1.2-megawatt fuel cell locomotive project.
- The New York Sheraton became the first Manhattan hotel to add fuel cells for supplemental power.
- Korea's Hyundai and Russia's Avtovaz showed fuel cell vehicles at the Geneva Auto Salon.
- Beijing's HYFORUM conference, organized by the German Forum für Zukunftsenergien (Forum for Future Energies), opened a window on China's hydrogen and fuel cell ambitions and potential.
- India unveiled its first fuel cell passenger car.
- BMW set speed records for hydrogen cars with internal combustion engines. An experimental three-wheeled car built by a small team of experts, HYSUN, covered the Berlin-to-Barcelona distance on 3 kilograms of hydrogen.

2005
- German fuel cell groups formed a national alliance and drafted a road map.
- General Motors launched the third-generation Sequel fuel cell vehicle at the Detroit Auto Show (figure 3.1).
- German industrial gas company Linde proposed an 1,100-mile hydrogen highway, including cost estimates. *Hydrogen highway* is a concept that has surfaced in recent years denoting a chain of hydrogen filling stations and other infrastructure along a road or highway to allow hydrogen-powered cars to refuel and travel.
- Global winds are sufficient to meet the world's total energy needs, reported a Stanford University study.
- A new U.N.-sponsored hydrogen center opened in Istanbul with an international conference.

2006
- French industrial gas company Air Liquide was selected to run the five-year EU-sponsored HYCHAIN-MINITRANS demonstration

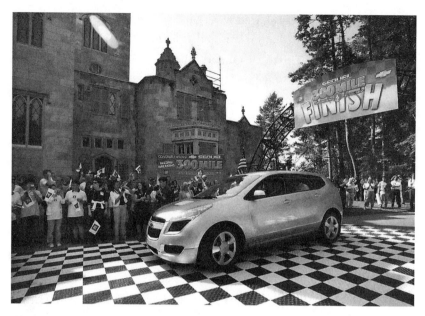

Figure 3.1
Cheered by well-wishers, one of two Chevrolet hybrid fuel cell Sequel SUVs arrives at the finish line in Tarrytown, New York's Lyndhurst Castle after completing a 300-mile drive on one tank of hydrogen without refueling from its starting point in upstate Honeoye Falls, New York.

project, deploying innovative low-power fuel cell vehicle fleets in four regions of Europe to initiate an early market for hydrogen as alternative fuel.

- General Electric developed a plastic-based electrolyzer for low-cost hydrogen production.
- California maker of hydrogen tanks and equipment Quantum won a contract to convert 15 hybrid internal combustion Prius cars to hydrogen for Norway.
- Big knowledge gaps and misperceptions about hydrogen persist, reported a DoE survey.
- The first American solar hydrogen house opened at the U.S. Merchant Marine Academy on Long Island (figure 3.2).
- The University of Minnesota launched a wind-to-hydrogen-to-fertilizer project.
- General Motors announced 100-plus fuel cell SUV fleet as a demonstration and learning project for ordinary drivers.

Figure 3.2
The banner says it all. "America's First Solar Hydrogen Home" was built on the campus of the U.S. Merchant Marine Academy that opened in Kings Point, Long Island, New York, in the summer of 2006. Electricity is produced by a 5 kilowatt PlugPower GenCore PEM fuel cell, and hydrogen is generated by a Proton Energy Systems HOGEN 40 RE PM electrolyzer.

- In Berlin, BMW turned reporters loose in the new Hydrogen-7 bifuel luxury sedans as a prelude to delivering more to hand-picked VIP customers in 2007.
- Ballard Power Systems signed a contract to deliver 2,900 fuel cells to General Hydrogen for forklifts and materials-handling equipment.
- Japan's Railway Technical Research Institute conducted the first-ever trials of a fuel cell–powered railway car.

2007

- General Motors unveiled its Volt as the first example of electric drive train architecture for both internal combustion engines and fuel cells. In its initial commercial 2011 version, the Volt was being sold exclusively as a internal combustion engine plug-in electric vehicle, but General Motors plans to replace the engines with a fuel cell eventually.

- The National Renewable Energy Laboratory cooperated with utility Excel Energy in starting an experimental wind-to-hydrogen energy system.
- An EU consortium started building a small fuel cell plane, Boeing planned fuel cell plane tests later in the year, and the European Space Agency funded studies for a future liquid hydrogen–fueled hypersonic transport plane.
- The seventieth anniversary of the Hindenburg disaster prompted an article series in the *New York Times,* and the German Hydrogen and Fuel Cell Association saw the diminishing of the *Hindenburg syndrome*—a term to describe irrational fears among those who believe hydrogen is an immensely dangerous substance.
- An Ohio State student team built a fuel cell racer to attempt a speed record on the Bonneville Flats in Utah.
- Europe's parliament in Strasbourg called for a green hydrogen economy.
- Two General Motors fuel cell Sequel SUVs covered 300 miles on one tank of hydrogen each.
- In Scandinavia, Iceland began to offer hydrogen rental cars and equipped a tourist ship with a fuel cell auxiliary power unit. Norway opened the first hydrogen station on its Hydrogen Highway.
- North Dakota opened an experimental wind-to-hydrogen station and showed a tractor partially fueled by hydrogen.
- A Ford fuel cell Fusion set a speed record of more than 200 mph at Bonneville.
- A Chinese company showed a hydrogen fuel cell bike at a Shanghai trade fair (figure 3.3).
- The European Space Agency funded studies toward a liquid hydrogen–fueled hypersonic transport airplane.

2008

- Abu Dhabi committed $15 billion to alternative energy, including a 420-megawatt natural gas-to-hydrogen power plant.
- BNSF Railway and Denver's Vehicle Projects developed a hybrid fuel cell switch locomotive.
- A hydrogen cell phone and recharger made their debut at the Las Vegas Consumer Electronics Show.

Figure 3.3
Chinese PEM fuel cell developer Shanghai Pearl Hydrogen Power Source Technology Co. showed its fuel cell bike, said to be China's first, at the 2007 Ninth China International Exhibition on Gas Technology, Equipment and Applications in Shanghai.

- Boeing successfully flew a small fuel cell airplane in Spain.
- Saying, "We can mass produce these now," Honda inaugurated the first small fuel cell car assembly plant in Japan.
- The port city of Hamburg received its first fuel cell river-and-lake tourism ship.
- IdaTech and Ballard Power Systems signed a large fuel cell deal with a major India telecom supplier.

2009

- BMW showed zero-carbon hydrogen electric cars at the U.N. Climate Summit in Poznan, Poland.
- A University of Illinois team developed the smallest PEM fuel cell yet (3 × 3 × 1 millimeters).
- Daimler announced plans to start small-series fuel cell production this year, and Toyota planned to launch a commercial fuel cell car in

2015. Ford and Chrysler dropped major parts of their hydrogen and fuel cell programs, mostly because of economic constraints caused by the global financial meltdown, although Ford was continuing some R&D work.

- Fiat's New Holland agriculture subsidiary rolled out first new hydrogen fuel cell tractor (Allis Chalmers had built one fifty years earlier).

- New DoE Secretary Steven Chu was described as interested in stationary fuel cell systems but skeptical about transportation and other hydrogen applications.

- Oak Ridge researchers were developing an enzyme "cocktail" that produces hydrogen from cellulosics and water at ambient temperatures.

- Shell announced scaled-back solar, wind, and hydrogen investments.

- BMW designed the most efficient (42 percent) hydrogen internal combustion engine yet.

- German chancellor Angela Merkel launched the first renewable hydrogen hybrid power plant.

- The DoE zeroed out transport and other hydrogen/fuel cell programs from the 2010 budget request, and former General Motors fuel cell chief Byron McCormick resigned in protest from the department's hydrogen and fuel cell technical advisory committee.

- Riversimple in the United Kingdom unveiled a small, low-power city hydrogen fuel cell car with hopes of changing the auto industry.

- Global reinsurer Munich Re launched an exploratory half-trillion-dollar solar Desertec initiative linking Europe, North Africa, and the Middle East.

- Mercedes launched small-series production of fuel cell cars, and General Motors showed its smallest fuel cell stacks yet.

- Thirteen Japanese energy companies banded together to push for the creation of a national hydrogen fueling infrastructure by 2015.

- In Germany, eight major carmakers signed a letter of understanding to push commercialization of fuel cell cars by 2015, and Germany launched national hydrogen fueling infrastructure plans.

- More than two dozen hydrogen cars were at the Copenhagen Climate Summit, but with little political impact.

2010

- Start-up Bloom Energy unveiled a mystery Bloom Box solid oxide fuel cell (SOFC) power plant in California.
- Toyota deployed ten fuel cell hybrids at JFK Airport at the start of its 100-vehicle U.S. demonstration fleet.
- Governor Arnold Schwarzenegger urged the National Hydrogen Association meeting to "wake up the federal government" to hydrogen.
- German aerospace agency DLR prepared for a transatlantic flight with a fuel cell plane.
- The National Hydrogen Association and U.S. Fuel Cell Council merged into the new Fuel Cell and Hydrogen Energy Association.
- A major study, coordinated by McKinsey & Company and based on data from thirty-one companies and organizations, was presented in November in Brussels calling for a Europe-wide market launch plan for fuel cell electric vehicles and hydrogen infrastructure and a coordinated rollout of battery-electric vehicles, plug-in hybrid electric vehicles, and a battery-charging infrastructure.

2011

- Mercedes-Benz sent three F-cell B-class cars on a 125-day trip around the world to demonstrate the market readiness of fuel cell vehicles.
- Boeing won a patent for an innovative ring-shaped liquid hydrogen tank for a future generation of blended wing-body airplanes.
- Europe's aerospace giant EADS (European Aeronautic Defence and Space Company), the parent of plane manufacturer Airbus, announced a concept study and plans to build an environmentally benign liquid hydrogen–powered hypersonic 100-passenger airliner at the Le Bourget Air Show in Paris.

4

Producing Hydrogen from Water, Natural Gas, and Green Plants

It was not an auspicious beginning for America's first operational solar hydrogen plant, and it certainly was not a big deal in usually sunny southern California. The morning sky over El Segundo on September 26, 1995, was overcast. A light drizzle fell on the 150-plus people—engineers, environmentalists, government officials, Xerox corporate executives, curious locals—who had gathered on a half-acre site wedged between the buildings of a corporate office park a couple of miles south of Los Angeles International Airport. They were there to witness the official start of the first operational solar hydrogen facility in the United States, making nonpolluting gaseous hydrogen fuel for three pickup trucks. These "sunlight-to-tailpipe" Ford Rangers cruised the streets of El Segundo and neighboring communities in the Los Angeles Basin under experimental permits issued by the California Air Resources Board.

Probably even more disappointing than the weather to James Provenzano, the executive director of Clean Air Now (CAN), the local environmental group that ran the project in partnership with the Xerox Corporation, was the meager media turnout. Although the prestigious *Los Angeles Times*, a suburban paper, and a few local radio stations gave the event some coverage, television reporters—on whom CAN had counted to give the project wide exposure—stayed away in droves. "That was the day of the closing arguments of the Simpson trial," Provenzano recalled ruefully later. "The jury went out the next day." Prime-time coverage of the trial of O. J. Simpson, football superstar turned Hollywood celebrity, for the alleged murder of his wife, had preempted just about anything else on local and national TV.

This $2.5 million solar hydrogen project had begun eighteen months earlier as a gleam in Provenzano's eyes, then the manager of environmental programs for Xerox in El Segundo, and Paul Staples, a local environmental activist. Both were working with CAN, founded and headed by the late Robert Zweig, a physician from nearby Riverside and a tireless hydrogen advocate for more than twenty years. Zweig, a long-time official of the American Lung Association, had access to data on the ravaging effects of air pollution on humans in the Los Angeles Basin and elsewhere. In 1973, around the time of the first oil crisis, Zweig had become convinced that one major way to drastically reduce pollution-caused illnesses such as emphysema, asthma, and lung cancer and their ballooning societal costs was to substitute a zero-emission fuel such as hydrogen for conventional carbon-based fuels.[1]

The CAN-Xerox facility incorporated, in embryonic form, the whole chain of equipment needed to make hydrogen from sun and water to run two pickup trucks for Xerox and one for the City of West Hollywood. Solar radiation, captured by a 48-kilowatt array of fresnel-lens-enhanced photovoltaic system, was turned into electrical energy, which was fed into an electrolyzer that produced about 1,500 to 2,000 standard cubic feet (scf) of hydrogen a day from deionized tap water in a separate feed-water preparation system. Next, a mist eliminator removed excess water vapor from the hydrogen gas, which was then piped to an interim gas holder (a metal tank) before being conveyed to a compressor, which compressed the gas to 5,000 pounds per square inch (psi). After that, the compressed gas was routed to a high-pressure dryer, which removed the last traces of water from the gas before it was fed into a high-pressure storage device holding about 13,000 scf of gaseous hydrogen, also at 5,000 psi. Because of the small capacity of the electrolyzer (this was only a demonstration project), but also to meet anticipated future needs for future hydrogen vehicles, additional hydrogen was trucked in periodically by the industrial gas maker Praxair from a production facility in Ontario, about 60 miles inland, replenishing a supplemental 80,000-scf, 2,200-psi storage system with hydrogen. Finally, both the supplemental storage system and the high-pressure tank fed fuel into the hydrogen "gas station," which then refueled the supercharged 1993 and 1994 Ford Rangers, which had been outfitted with lightweight pressurized gas tanks, each holding about 2,600 scf of hydrogen gas at 3,600 psi—equal in energy content to about 5 gallons of gasoline and good for about 140 miles. In regular service, each truck covered about 100 miles a week.

The Xerox trucks carried maintenance equipment and tools; the West Hollywood truck was used for environmental audits and to haul recycling bins.

The project came to an end in 1997 when Xerox decided it needed the real estate for other purposes. The storage tanks and the liquid-hydrogen pump stayed at the site, permitting the refueling of liquid-hydrogen-powered trucks operated by the company; however, the solar panels, the electrolyzer, and related equipment were moved some 100 miles east to Palm Desert, at the edge of the Mojave Desert, where they became part of a new hydrogen fuel system built by a bus company, for golf carts converted to fuel cell power (golf carts are street legal in some resort towns), and eventually for hydrogen fuel cell buses operated as part of the California Fuel Cell Partnership.

By 2009, solar-generated hydrogen was among the major topics discussed at the Fiftteenth Solar PACES (Power and Chemical Energy Systems) conference in September in Berlin. Solar PACES is an international agreement launched three decades earlier by the International Energy Agency; perhaps as a sign of increasing urgency or growing interest, or both, this conference series, which used to be held biannually, shifted to an annual cycle beginning in 2009. Almost two dozen of the roughly 200 papers presented to the 700 participants dealt with solar hydrogen production and solar water splitting, including policy and marketing aspects. Arguably the most interesting presentation dealt with the successful demonstration and tests the previous year of a 100-kilowatt solar tower HYDROSOL pilot facility, part of the Plataforma Solar de Almeria solar power plant in Spain.[2] The paper, a collaboration of almost two dozen researchers from five countries, reported on the process: a two-chamber system designed by the German aerospace agency DLR and first tested five years earlier in a DLR solar furnace at a research center outside Cologne. The actual water-splitting step—splitting steam, really—takes place at 800°C; a second step, in which the system is flushed with nitrogen to release the oxygen from the metal oxide redox system created in the water-splitting step, requires higher temperatures of 1,200°C. The bottom line, wrote the authors, was that "significant concentrations of hydrogen were produced with conversion of steam of up to 30%." Other papers came from Israel's Weizmann Institute of Science; France's Promes-CNRS Laboratory; Science

Applications International Corp. in San Diego, California; the Florida Solar Energy Center in Cocoa; Sandia National Laboratories; Tokyo Tech in collaboration with Australia's research organization CSIRO; and others.

In between that 1995 El Segundo CAN-Xerox project and the much bigger, more sophisticated, and more advanced 2009 Almeria solar hydrogen production R&D effort, a number of other solar-powered water-splitting R&D came and went, none of them leading to commercially viable, economical hydrogen production. That goal has not yet been achieved.

Early Examples: Electrolysis in the Desert

Solar-powered water electrolysis to make hydrogen fuel was also the basic operating principle of two German facilities: Solar-Wasserstoff Bayern (Solar Hydrogen Bavaria) and HYSOLAR, a joint German-Saudi operation, with equipment churning out solar hydrogen near Saudi Arabia's capital of Riyadh and in a parallel but smaller research-and-teaching setup on the outskirts of Stuttgart. Solar-Wasserstoff Bayern was jointly operated by Bayernwerk, Bavaria's leading utility (renamed E.ON Energie AG in 2000 as part of a merger); carmaker BMW; industrial gas supplier Linde; and the electrical equipment conglomerate Siemens. Launched in 1986, the 12.4-acre facility was nestled in an idyllic valley near the small community of Neunburg vorm Wald, close to the border with the Czech Republic. It was intended as a permanent testing facility for whatever new type of hydrogen technology came off the drawing boards. Powered by arrays of various types of photovoltaic panels rated at close to 350 kilowatts, the complex housed advanced low-pressure water electrolyzers; systems for compressing, purifying, drying, and storing hydrogen and oxygen; gas-fired boilers for testing various mixtures of natural gas and hydrogen; several types of fuel cells; utility and auxiliary systems; and an experimental liquid-hydrogen filling station for cars that eventually evolved into a fully automatic robotic system. Solar-Wasserstoff Bayern operated for more than ten years at a total cost of about $89 million. It was closed in 2000 because of economic realities and the unwillingness of the sponsors to fork out more funds for something that seemed to be too far in the future.

HYSOLAR, the result of a cooperation agreement signed in 1986 between Saudi Arabia and Germany, was created to investigate solar hydrogen production and utilization. The Saudi team included researchers from the King Abdulaziz City for Science and Technology in Riyadh and three other universities. DLR, the German Aerospace Center, ran the German side. During the first eight years, HYSOLAR's total costs were about 64 million deutsche marks (about $37 million), split evenly between Saudi Arabia and Germany. After some eight years of preparation and testing, some false starts, and some delays, HYSOLAR officially began churning out hydrogen from solar electricity in early 1994. Power was produced by an array of concentrator photovoltaic panels initially rated at 350 kilowatts. HYSOLAR's R&D program covered most aspects of solar hydrogen technology, including catalytic combustion, hydrogen internal combustion engines, and fuel cell technology. Fundamental research was being done on photoelectrochemistry, electrochemical energy conversion, and combustion processes, in addition to hydrogen energy systems. The program ran for ten years, from 1986 to 1995, at a total cost of about 83.5 million deutsche marks (roughly, $52 million), but various parts were extended past the official end date.

Hydrogen Production: Count the Ways

Today, just beginning the second decade of the twenty-first century and decades after research of hydrogen as fuel got underway in earnest, no single or best technology has emerged as the clear winner, and research continues more intensively than ever before. The industrial standby of hydrogen production, steam reforming of natural gas, continues to be the most economical method of choice widely used in the petrochemical and other industries, and there have been significant improvements over the years. But most hydrogen supporters regard it as an interim or bridge production technology on the road to renewable energy–based methods. Electrolysis, the classical method of splitting water into hydrogen and oxygen by running a current through it, is used in niche applications and where very low-cost electrical energy is available from hydroelectric plants, for example. But new approaches scarcely imagined a couple of decades ago have been sprouting in recent years.

A case in point was the U.S. Energy Department's May 2009 Hydrogen Program Annual Merit Review in Arlington, Virginia, an annual rite that helps determine who gets funded by the department in coming years and attended that year by almost 1,300 scientists, government officials, and industry people.[3] Topics of the production and delivery sessions alone—other sessions dealt with storage, fuel cells, manufacturing, systems analysis, and the vehicle technologies program (which included education, safety codes and standards, advanced combustion, fuel technologies and others)—covered distributed bio-derived liquids production, high-temperature thermochemical nuclear hydrogen, biological electrolysis, photoelectrochemical hydrogen production, biomass gasification, and hydrogen from coal. The reviewers listened to thirty-four production presentations and another sixteen on delivery, and outside the meeting rooms there were an additional twenty-two posters on production and seven on delivery. Among the paper titles were these:

- "Hydrogen Generation from Biomass-Derived Carbohydrates via the Aqueous-Phase Reforming Process"
- "Solar Thermal Ferrite-Based Water Splitting Cycles"
- "Maximizing Light Utilization Efficiency and Hydrogen Production in Microalgal Cultures"
- "PEM Electrolyzer Incorporating an Advanced Low Cost Membrane"
- "A Novel Slurry Based Biomass Reforming Process"
- "Scale-Up of Hydrogen Transport Membranes for IGCC (Integrated Gasification Combined Cycle) and FutureGen Plants" in the Hydrogen Production from Coal category[4]

A similar range of production technologies, organized in thirteen subcategories, were in evidence at the May 2010 Eighteenth World Hydrogen Energy Conference in Essen. They covered a variety of topics—for example:

- Photobiological and fermentative hydrogen production: "Photohydrogen Production from Thermophilic Aerobic Digestion Effluent and Distillery Wastewater" from Chung Hsing University Taiwan
- Thermochemical cycles: "Two-Step Water Splitting by Cerium Oxide-Based Redox Pairs," from the German Aerospace Research Center

• Hydrogen from renewable electricity: "Study of Hydrogen Production from Wind Power in Algeria," from the Centre de Developpement des Energies Renouvables in Bouzareah, Algeria).

The Holy Grail: Extracting Hydrogen from Water

Arguably the great attraction of hydrogen as a fuel has to do with the fact that you can produce it from something as mundane as water. The idea has a—do I dare say it? – primordial quality with a whiff of the eternal and the universe. But breaking up the water molecule is a difficult, energy-intensive business because of the strong chemical binding forces between hydrogen and oxygen. Hydrogen ordinarily does not exist in a free state; it almost always occurs in a compound of some sort. There are many combinations other than the H_2O molecule. Hydrogen gas usually exists only in the molecular state, H_2.

Water can be broken up in many ways, including directly, with extreme heat; with the help of chemicals in two or more steps; with a combination of heat and chemicals; by the action of certain microorganisms found in the oceans; and by running an electrical current through water (or even steam) in electrolysis, a process familiar to high school students everywhere.

Electrolysis is a proven method for making hydrogen and oxygen on an industrial scale. However, it has been used to a significant extent only in places where electricity is very cheap, such as Canada and Norway, with their vast hydropower resources. For a long time, making hydrogen the electrolytic way was considered economically justifiable only when it was to be used as a high-value chemical feedstock and only when high purity was required. In the late 1960s and the early 1970s, many researchers believed that electrolysis was almost hopelessly outdated and inefficient as a method of making something that would serve, after all, only as fuel to be burned again. But HYSOLAR, SWB, and CAN were clear indications that electrolysis was far from written off. The U.S. Department of Energy's 2009 review meeting agenda included advanced electrolysis concepts such as proton exchange membrane (PEM) electrolyzers, conceptually the reverse of PEM fuel cells used by most car manufacturers and pursued by Giner Electrochemical Systems in Newton,

Massachusetts, and high-pressure electrolysis under development at Avalence in Milford, Connecticut (more on these concepts later).

In principle, electrolysis of water is very simple: Two electrodes, one positive and one negative, are immersed in water that has been made more conductive by the addition of an electrolyte, either acidic or basic. When direct current power is applied, hydrogen begins to bubble up at the negatively charged electrode (the cathode), and oxygen rises out of the solution at the positively charged electrode (the anode). Fresh water of high purity—important because otherwise the accumulating salts would clog up the electrolyzer—is continuously fed to replace the water that has broken down into hydrogen and oxygen.

An industrial electrolysis plant is considerably more complex. In addition to the basic tank arrangement—electrodes, electrolyte, separators to keep the electrodes from touching and shorting out and to separate the two gases at the source, and a container, all of which can take various configurations—such a plant usually requires the following:

- An electric power converter that changes conventional alternating current to direct current at a small but significant loss
- Equipment to distribute electric power to the electrodes
- A system of pipes to carry oxygen and hydrogen away from the cells
- A special separation system (in some plants) to remove the gases from the electrolyte
- Cooling machinery to remove heat generated during electrolysis
- Drying equipment to dry the gases after they have been separated from the electrolyte

Commercial electrolysis cells date back to the 1890s. However, electrolysis has never been a major industrial-scale activity in the United States, where heavy users of hydrogen, such as producers of fertilizers and the space program, have always relied on cheaper hydrogen made from steam-reformed natural gas. "Electrolysis has traditionally been considered one of the more expensive methods of hydrogen production, and electrolyzers have been assumed to be inefficient and expensive," according to a 1975 survey of hydrogen production and utilization methods by the Institute of Gas Technology.[5] This is not quite the case, though, the report continued. "On the contrary, it is the electric generation step that is expensive and inefficient. Most commercial electrolyzers available today are capable of electricity-to-hydrogen efficiencies above 75 percent, while their capital-cost potential is far less than that of the power stations that would be required to run them."

Today many commercial electrolyzers operate efficiencies as high as 80 to 85 percent (of higher heating value; both higher and lower heating values are used to describe efficiencies, leading occasionally to confusion in competing claims), and 90 percent has been demonstrated in the laboratory.

Compared to proposed water-splitting methods such as direct solar water splitting (which requires extremely high temperatures) and thermochemical, electrochemical, or photobiological hydrogen production techniques (which require sophisticated handling of biological or chemical materials), relatively simple electrolyzers have advantages, although newer designs presumably are getting more complex as well. Since there are no moving parts, they operate trouble free and more or less automatically, requiring servicing, such as the exchange of corroded electrodes, only every few years.

Electrolysis also lends itself to operation under higher pressures, which aids process efficiency. Importantly, electrolysis produces pure hydrogen gas. Fossil-fuel-derived hydrogen is usually somewhat contaminated—potentially problematic for its use in fuel cells. It can be cleaned up, but that adds more steps and therefore increases cost.

The amount of energy needed to decompose water into hydrogen and oxygen by electrolysis is, in theory, exactly the amount of energy given off in the reverse process, in which hydrogen burns and recombines with oxygen into water vapor. In practice, there are losses in both electrolyzers and fuel cells; it takes more energy to split water than can be retrieved by combining the resulting hydrogen and oxygen in a fuel cell. A completely efficient electrolysis cell would require 94 kilowatt-hours to make 1,000 cubic feet of gaseous hydrogen. Not all the energy need be supplied as expensive electricity, only 79 kilowatt-hours. The rest can be brought into the process as simple heat, a less sophisticated and less costly form of energy—an approach that has periodically stirred interest in the idea of electrolyzing steam.

Electrolyzers

There are essentially two types of industrial electrolyzers: the unipolar (tank-type) electrolyzer and the bipolar (filter-press) types. Though there have been improvements in materials, design, and conversion efficiency (from about 70 to 75 percent in the 1970s to about 80 to 90 percent now, based on the higher heating value of hydrogen), the basic concept has not changed in about seventy years.[6]

An early landmark in electrolysis technology was the decision by a Norwegian utility, Norsk Hydro, to use an electrolyzer to make synthetic fertilizer. One of the first truly big electrolyzers, it was built in 1927. The first large electrolyzer in North America was built around 1940 in Trail, British Columbia, by the mining company Cominco. After 1945, many large electrolyzers were built around the world. The three largest (all used in the manufacture of fertilizer) are a plant in Nangal, India, built by De Nora of Milan, Italy; the Norsk-Hydro plant at Rjukan; and a plant at the Aswan Dam in Egypt, erected by the German company Demag.

The unipolar electrolyzer is the older and simpler type. Alternate electrodes are hung in a tank filled with electrolyte. The electrodes are separated from adjacent opposite-charge electrodes by a diaphragm, usually of asbestos, that allows passage of the electrolyte but prevents mixing of the hydrogen and oxygen gases. The unipolar electrolyzer has a humber of main advantages: it requires a relatively small number of parts, the parts are inexpensive, and an individual cell can be shut down for repair or replacement simply by short-circuiting two adjacent cells while the rest of the cells continue making hydrogen. The main disadvantages are that it is unsuitable for high-temperature operation because of heat losses due to large surfaces, and it usually requires more floor space than a bipolar electrolyzer does.

In a bipolar electrolyzer, each electrode has both a positive and a negative face, with the positive face in one cell and the negative face in the adjacent cell. Proponents say that bipolar electrolyzers take up less floor space than unipolar ones and are better suited to more efficient high-pressure and high-temperature operation. Their drawbacks are that they require much more precise tolerances in construction and are more difficult to maintain. For example, if one cell fails, the entire assembly has to be shut down.

One of the best-known commercial examples of the unipolar cell in North America was the range of so-called Stuart cells, made for decades by the Electrolyser Corporation in Toronto. In the mid-1990s Electrolyser revived the idea of home production of hydrogen for cars. Andrew Stuart, presenting the concept in May 1997 at the Canadian Hydrogen Workshop in Toronto, said the company was developing a washer-dryer-size electrolysis appliance with a built-in compressor that would produce hydrogen overnight. The aim was to produce the appliance for about U.S.$1,500 (if manufactured in volumes of 1 million units or more and sold with the car) or for about U.S.$5,000 with a production volume of about 10,000. A first prototype of this personal fuel appliance was demonstrated to Ford Motor Company experts in autumn 1999 and shown at the spring 2000 National Hydrogen Association conference in Vienna, Virginia. But Stuart ceased operations in early 2005 when it was acquired by another Canadian fuel cell developer, Hydrogenics of Mississauga, Ontario.

Teledyne Energy Systems, located near Baltimore, continues to be among the preeminent makers of bipolar electrolyzers. Its electrolyzers range from small units for laboratory purposes to big machines capable of generating several hundred pounds of hydrogen per day. Teledyne got into the business in 1971 when it acquired fuel cell and electrolyzer technology from Allis-Chalmers, which had developed such know-how mainly while working as a contractor for the U.S. space program. Teledyne has been credited with being one of the few companies in the United States that continue to invest time, money, and other resources into improving standard electrolyzer design, and it is a major player in this sector.

Originally spearheaded by General Electric in the 1970s for commercial hydrogen production and generating oxygen in closed environments such as submarines, solid polymer electrolyte (SPE) technology is another example of advanced electrolysis.[7] Analogous to the proton exchange membrane (PEM) technology used in fuel cells, SPE electrolysis replaces the conventional liquid electrolyte with a solid sheet that looks and feels somewhat like Teflon. When soaked in water, it becomes an excellent conductor. Unlike standard electrolyzers, it does not require the addition of acid or alkalis to the water to help drive the electrolysis process. Eventually GE sold both PEM electrolysis and the PEM fuel cell technology to Hamilton Standard, a subsidiary of United Technologies Corporation, which led to the creation of International Fuel Cells as a subsidiary of Hamilton Standard. Over the years, some of GE's core personnel left the company, dispersing PEM know-how to new players that today are involved in the intense international competition to bring PEM technology to the market.

PEM electrolysis moved to the fore again with the creation of a new company, Proton Energy Systems of Rocky Hill, Connecticut, to commercialize the technology initially developed by Hamilton Standard. Started in 1996 by a small group of former Hamilton Standard employees, Proton Energy has developed a series of SPE/PEM electrolyzers for industrial use and has also developed fuel cells, including a regenerative type that can function reversibly as an electrolyzer and a fuel cell.

The conversion efficiency of PEM electrolysis is about the same as that of conventional electrolyzers with liquid electrolytes—about 80 to 90 percent. But PEM electrolysis is able to generate hydrogen at high pressure without the need for an extra compressor and with high purity (more than 99.9999 percent). PEM electrolyzers are notably compact, and, like PEM fuel cells, they are claimed to take rapid increases or decreases in electrical input in stride. Their proponents say that electrolyzers and fuel cells with liquid electrolytes do not respond readily to such changes: too much power input too quickly can make the liquid electrolyte bubble rather than split water into hydrogen and oxygen or generate electricity.

Splitting Steam

Another variant of electrolysis technology was the development of high-temperature steam electrolysis, initially by General Electric and later by Germany's Dornier System. In the late 1960s, General Electric began experimenting with electrolyzing 1,000°C (1,832°F) steam with an electric current. The main advantage claimed for this type of electrolysis was that at high temperatures, the need for expensive, high-quality electricity is drastically reduced because much of the energy would be provided as raw heat. The hope was that the overall cost of producing hydrogen from water would fall significantly with this approach.

Carlsbad City Library - Cole
CHECKOUT RECEIPT

Title: Archenemy
[videorecording]
Item ID: 31245011652579
Date due: 5/28/2021,23:59

Title: We got this
[videorecording]
Item ID: 31245012144170
Date due: 5/28/2021,23:59

Title: Tomorrow's energy :
hydrogen, fuel cells, and the

Item ID: 31245006638336
Date due: 5/28/2021,23:59

--PLEASE KEEP THIS
RECEIPT--
To renew, please call:
760-434-2870
or visit our website at:
www.carlsbadlibrary.org
Thank You!

Dornier System began looking into high-temperature steam electrolysis in 1975. The program lasted about fifteen years, achieved efficiencies of about 93 percent, and reduced electricity consumption by about 30 to 45 percent relative to conventional electrolysis. The core of the system was a solid ceramic oxide electrolyte consisting of zirconium oxide and coated with porous electrodes with a high degree of conductivity for oxygen ions at operating temperatures, with hydrogen evolving on the cathode side and oxygen on the anode side. But as the shock waves generated by the oil crises ebbed away and gasoline became cheap again, the program faded into oblivion.

But the U.S. Department of Energy decided to take another look at the technology again, embarking on a long-term high-temperature electrolysis r&d project in 2003 that was supposed to culminate in a 1-megawatt engineering demonstration plant in 2015. Led by researchers at Idaho National Laboratory, project partners include the Massachusetts Institute of Technology; Ceramatec, a developer of solid-state and aqueous ionic-conducting ceramic materials and fuel cells, based in Salt Lake City, Utah; France's St. Gobain Ceramics, a maker of high-temperature refractory materials; NASA–Glenn Research Center; Argonne National Laboratory; and Virginia Tech and Georgia Tech. With about $2.5 million in funding in fiscal 2009, plans for that year included expansion of the testing laboratory to permit parallel testing of up to five experiments, tests of various cells and stacks from several vendors, examine solid oxide fuel cell degradation issues, explore advanced electrode materials, experiment with steam concentration within cells, and other tasks.

Thermochemical Water Splitting

Thermochemical water splitting—the idea of combining high temperatures with the high chemical reactivity of certain continuously recycled chemicals to break up the water molecule into its components hydrogen and oxygen—was a hot field of hydrogen research in the 1970s, when it first became clear the West would have to wean itself from its dependence on Middle Eastern oil. It revived a technology that had originated fifty years earlier as an effort to split out hydrogen from water so as to make ammonia for fertilizer. One of the earliest proposals for thermochemically producing hydrogen from water was patented in 1924 in England by a scientist working for a company called Synthetic Ammonia and Nitrates, Emil Collett, who postulated a two-step cycle. It probably would not have worked very well; thirty years later, researchers decided that for a number of theoretical reasons, at least three cycles were needed.

In the United States, splitting water in a heat-assisted, closed-cycle chemical process to make hydrogen fuel was first explored in the early 1960s, when an investigation of "chemonuclear" processes was begun at Brookhaven National Laboratory. The idea was to use a large nuclear power plant to produce both hydrogen and ammonia to meet the then-skyrocketing demand for fertilizer. In 1966, building on the idea, two General Motors scientists, James Funk and Robert Reinstrom, wrote a fundamental paper that still crops up in the literature as one of the hydrogen economy's basic documents.[8] Funk and Reinstrom were then working in Indianapolis for GM's Allison Division, which was building tanks for the U.S. Army. The army was investigating the idea of what it called a "nuclear-powered energy depot" to make hydrogen as tank fuel. The idea, militarily a dud, did not look very promising scientifically. "It appears unlikely," Funk and Reinstrom concluded, "that a compound exists, or can be synthesized which will yield a two-step chemical process superior to water electrolysis."

Interest in thermochemistry waned in the 1980s, and since then it has been largely relegated to the sidelines. In the 1990s, General Atomics of San Diego was among the few stalwarts still promoting the concept in the context of applications for a high-temperature nuclear reactor.

With the new century, interest in high-temperature thermochemical hydrogen production revived, in part because of the renewed interest in nuclear energy and also high-temperature solar energy production—again, the result of the growing awareness of the need for clean energy, and more energy in general, as energy consumption everywhere was projected to climb dramatically. At the 2009 DoE Hydrogen Program review, seven presentations from researchers at organizations such as the Florida Solar Center, General Atomics, the University of Colorado, Argonne National Laboratory, Sandia National Laboratories, and Savannah River National Laboratory discussed the solar cadmium-hydrogen production cycle, solar-thermal ferrite-based water-splitting cycles, sulfur-iodine thermochemical cycles, and other processes.

Producing Hydrogen from Fossil Fuels and by Other Means

Many hydrogen proponents are generally in agreement that the initial impetus to get hydrogen going in the commercial marketplace would have to come from the use of carbon-containing fuels, such as natural gas, biomass, and maybe even coal.

In 2006, about 47.8 million metric tons of hydrogen were produced commercially per year around the world.[9] This may sound like a lot, but it is not; it is only 1 or 2 percent of the world's energy demand. U.S. hydrogen consumption that year was 3,014 billion standard cubic feet (scf) of hydrogen, most of it—1,967 billion scf—in refinery operations, followed by ammonia production (693 billion scf), methanol (60 billion scf), and the category "other" (294 billion scf).

Hydrogen is stripped out of hydrocarbon fuels, typically natural gas, in a method called steam reforming, in which natural gas reacts with steam at about 1,500°F to 1,600°F with the help of a nickel catalyst. The result is a mixture of hydrogen, carbon monoxide, carbon dioxide, steam, and unreacted methane. This mixture is cooled down to about 750°F and reacted further with more steam over a water gas shift catalyst, producing more hydrogen and converting the carbon monoxide to carbon dioxide. The CO_2 and other impurities are removed by pressure swing adsorption, a process that leaves pure hydrogen. A "sorption-enhanced" process developed by Air Products and Chemicals in the 1990s promised to reduce manufacturing costs by another 25 to 30 percent by removing CO_2 during the reaction.

Another advanced version of steam reforming suitable for producing high-purity hydrogen has been developed by H2Gen Innovations, a small company, in Alexandria, Virginia. Started in 2001 by Sandy Thomas, a former legislative assistant to Senator Tom Harkin (D, Iowa), and Frank Lomax as cofounders and co-inventors, H2Gen's proprietary process is essentially a more sophisticated version of steam reforming that produces hydrogen at what the company said are costs 25 to 50 percent less than existing conventional methods. The key innovations were a unified vessel design that combines two steps—reacting methane with steam and combining resulting carbon monoxide with water producing more hydrogen, plus oxygen—in one vessel; a greatly simplified low-cost pressure swing adsorption device for purification; a patent-pending catalyst that is sulfur tolerant; and other refinements. In a 2007 presentation, Lomax presented cost projections for three sizes of their machines capable of generating 113 kilograms per day, 567 kilograms per day, and 1,500 kilograms per day, respectively, at costs within shouting distance of 2009 gasoline prices (1 kilogram of hydrogen has roughly the same energy content of 1 gallon of gasoline): $5.42 per day for the smallest model, $3.10 for the midsized one, and $2.52 for the biggest one, which was then still under development.[10]

H2Gen Innovations has sold its hydrogen generators to France's Areva Group to help clean up and process used nuclear weapons by-product materials at U.S. DoE nuclear facilities, but also to, for example, Chevron for installation at a hydrogen fueling station near the Orlando, Florida, airport; for testing to Tokyo Gas; and for metal treatment and bell annealing, including to one customer in Cairo, Egypt. In 2007, the company's work was honored with a DoE Hydrogen Program R&D Award during the program's annual peer review.

A novel method of producing hydrogen from natural gas that surfaced around 1994 was the Kvaerner process, named after the major Norwegian gas and oil engineering and management company, Kvaerner Engineering, that developed it. (Norway has lots of natural gas as a by-product of its offshore oil fields.) The Kvaerner process involves adapting a plasma torch (on which Kvaerner had been working since 1982) in the decomposition of hazardous wastes. The hot plasma would be used to decompose natural gas into hydrogen and commercially salable carbon black, used in the rubber, tire, plastics, paint, and ink industries, in some new metallurgical industries as a reduction material, and a carbon additive or carburizer in steel and foundry operations. The process is claimed to result in almost 100 percent conversion of natural gas into hydrogen and carbon black.

Partial oxidation, conceptually similar to the Kvaerner process, uses part of the fossil-fuel feedstock to generate the requisite process heat, yielding only 20 to 60 percent carbon black; the rest of the combustion products goes up the smokestack as CO_2, nitrogen oxides, SO_2, and other emissions. Partial oxidation technologies gained ground in the 1990s, often in the context of developing onboard fuel reformers for cars and buses, but onboard reforming has been pretty much discarded as too cumbersome and complicated. An incomplete list of early developers of such reformers included Argonne National Laboratory, Epyx Corporation, International Fuel Cells, Johnson Matthey, Northwest Power Systems, and Shell.[11] More recently, partial oxidation was part of a General Electric project, with Argonne National Laboratory and the University of Minnesota as subcontractors, that looked at combining catalytic partial oxidation with steam methane reforming and water gas shift into one unit. That effort was completed in late 2008.

Low-temperature partial oxidation, along with catalytic autothermal reforming, is a key component in a process being investigated by the National Renewable Energy Laboratory (NREL) and the Coloroado School of Mines. NREL's lead researcher, Stefan Czernik, presented the project at the 2009 DoE Hydrogen Program Review meeting, "Distributed Bio-Oil Reforming." Begun in 2005, it is designed to produce high yields of liquid bio-oils, which can be stored and shipped to a site for renewable hydrogen production. The target is to produce hydrogen at costs of $3.80 per gallon gasoline equivalent with 72 percent energy efficiency in the conversion of bio-oil to hydrogen. Construction of a prototype system is expected for fiscal year 2011, and long-duration runs to validate the process are planned to get underway in 2012.

In the 1970s, in the aftermath of the oil shocks, a number of experts and organizations argued in favor of hydrogen production from coal via coal gasification—something the U.S. DoE began looking at again in the 1990s and is now actively supporting. Hydrogen production from coal is basically a partial oxidation process similar to that for heavy oil, explained an early study by the Jet Propulsion Laboratory: "The process is complicated by the necessity to handle a relatively unreactive fuel as a solid and to remove large amounts of ash," the study said. "The solids-handling problem has a severe impact on costs and prevents much of the technology and equipment developed for petroleum from being used in the conversion of coal. Coal, steam, and oxygen react in the basic gasifier processes to produce hydrogen."[12]

Partial oxidation is the hydrogen-producing step in what in the new century's first decade was perhaps the most ambitious, challenging effort to produce hydrogen from what is fast becoming the most contentious, most severely criticized, but also most abundant, fossil fuel, coal: the FutureGen project, the design of an ultraclean coal-burning power plant, supported by the DoE. The basic idea is to build a 275-megawatt plant in Mattoon, Illinois, that will combine ultra-high-efficiency dual-turbine integrated gasification combined cycle technology with carbon capture and storage. The primary turbine would be fired by hydrogen produced in the gasification step after carbon dioxide has been removed, and the hot exhaust gases would drive a second turbine for more power. The original 2003 concept, as developed by a nonprofit group of fourteen of the world's biggest coal producers, the Washington-based FutureGen Alliance (some have dropped out since then), also envisioned producing hydrogen as a by-product for use as a power source for car and bus fuel cells, but starting in 2009, that began to look doubtful. Part of the problem has been wildly escalating cost estimates, from the original $950 million to as much as $2.4 billion, with the DoE reluctant to spend more than the $1.073 billion it was prepared to contribute and the coal producer equally reluctant apparently to kick in more money. At one point, DoE reportedly pulled the plug on the project, but then relented and asked for a redesign, apparently without the hydrogen cogeneration part for transportation fuel. In 2010, the project, relabeled FutureGen 2.0, moved from Mattoon to a smaller 200-megawatt coal-fired plant in Meredosia, Illinois, where design and engineering studies got underway that fall.

Not only hydrogen but also many other low-polluting alternative fuels, such as methanol and ethanol but also biodiesel, synthetic gasoline, and new types of jet fuels, can be manufactured from many carbon-bearing sources, including coal and biomass. This is in fact what happened during the first ten years of the new century. During the second half of the decade, hydrogen in large measure disappeared from public view, especially in the United States, as the cleanest chemical fuel. It was replaced by the media and in the general public's perception by these other carbon-based liquid fuels that various interests touted as just about as clean, but much faster to introduce, more economical, and more convenient that would not upset existing fuel delivery infrastructures and still deliver on the clean-environment front.

In the pecking order of ease in making hydrogen from fossil fuels, methane, the main ingredient in natural gas, occupies the top spot because it usually has been cheap. It is easier to handle industrially than a liquid (such as crude oil) or a solid (such as oil shale or coal). It also contains the most hydrogen: four hydrogen atoms for each carbon atom. Next comes petroleum, with a hydrogen-to-carbon ratio of 1.5 to 1.6. Oil shale has a ratio of 1.6, but it is solid and therefore more difficult to handle than petroleum. Coal, posing the greatest difficulties for hydrogen extraction, has ratios ranging from 0.72 to 0.92.

Beyond the conventional, traditional production processes are ideas, concepts, experiments, and programs that attempt to produce hydrogen through imaginatively novel approaches. Some of them are inspired by energy-conversion processes found in nature.

One of the more successful new-technology startups is Virent Energy Systems in Madison, Wisconsin. Launched shortly after the turn of the century, it seeks to capitalize on concepts developed originally by Randy D. Cortwright and James A. Dumesic, scientists at the University of Wisconsin. The basic idea is to extract hydrogen using a platinum-based catalyst by breaking down liquefied carbon-neutral biomass materials such as glucose—corn syrup and wood pulp—but also leftovers from cheese production at relatively low temperatures of around 200°C into hydrogen gas, carbon dioxide, and methane. A report in *Nature* said that biomass is typically broken down by bacteria and fermentation, but that can be complex, inefficient, and expensive on an industrial scale.[13] Doing it in the liquid phase makes the whole process easier than other attempts that tried to do the same thing in a gaseous phase. In 2007, Virent signed a five-year agreement with Shell Hydrogen to develop and commercialize Virent's BioForming technology for hydrogen production.[14]

Researchers in the United States, Japan, and Europe, traditionally in Russia, and increasingly in other parts of the world—most prominently perhaps China and India, but also including Australia, Brazil, Iceland, Korea, New Zealand, Singapore, and Turkey—are investigating processes that use the direct reaction of sunlight with catalysts or biological organisms (such as blue algae). These processes split the water molecule with sunlight and are aided by a catalyst or semiconductors—the list of plausible and not-so-plausible hydrogen-making processes that have been suggested or tried goes on and on—and it goes back a long time. As early as the mid-1970s, one company, KMS Fusion of Ann Arbor, Michigan, had even investigated the use of high-powered lasers to initiate thermonuclear fusion and tried to show that hydrogen could have been produced in that fusion reaction.[15]

One of the first successful photolysis experiments, and a milestone in the history of advanced hydrogen generation, was reported in 1972 by two Japanese scientists, Akira Fujishima of Kanagawa University and Kenichi Honda of the University of Tokyo. They had managed to produce small amounts of hydrogen by shining light on an electrolysis system that used semiconductor electrodes powered by sunlight only, without any additional electric voltage. In their cell, a titanium dioxide (TiO_2) electrode was connected to a platinum black electrode through an external circuit. When the TiO_2 electrode was exposed to light, a current flowed from the platinum electrode back to the (TiO_2) electrode through the outside circuit, and Fujishima and Honda observed gas bubbling up from the electrolyte. From the direction of the current, they concluded that water can be decomposed by visible light into oxygen and hydrogen without the application of any external voltage and that hydrogen was produced at the platinum black electrode and oxygen at the TiO_2 end in minuscule amounts—around 1 percent.

Titanium dioxide is also a key ingredient in an electrochemical photovoltaic system that Michael Graetzel of the Swiss Federal Institute for Technology has been working on since the early 1990s. Dubbed originally "artificial leaf," the system of large panels and modules mimics natural photosynthesis, but instead of plant matter, it produces electricity and hydrogen. In Graetzel's system, TiO_2, an environmentally benign, semiconductive, low-cost ceramic material widely used in toothpaste and as a white pigment in paints, takes the place of expensive high-purity silicon. In initial tests, Graetzel claimed conversion efficiencies of 7 percent in direct sunlight and 11 percent in diffused light. At the time, he claimed that the artificial leaf may turn out to be the least expensive photovoltaic device yet, producing electricity for about $0.05 per peak watt.[16] In a 1996 report, Graetzel characterized his "chemical solar cell" as a molecular photovoltaic system based on the sensitization of nanocrystalline films by transition-metal charge-transfer sensitizers. The overall efficiency of converting solar energy into electricity was said to be in excess of 10 percent. In a 2008 joint article in *Nature Materials,* Graetzel and colleagues from the Chinese Academy of Sciences State Key Laboratory of Polymer Physics and Chemistry of the Changchun Institute of Applied Chemistry said they achieved cell efficiencies of 8.2 percent using a new solvent-free liquid redox electrolyte consisting of a melt of three salts.[17] These more recent projects were intended to substitute for conventional inorganic photovoltaic devices as electricity producers rather than to generate hydrogen. Still, Graetzel has continued with his research into solar hydrogen generation: In a February 2010 e-mail to me, he wrote, "We have always maintained a significant research effort in the field of solar hydrogen generation from water by photoelectrochemical cells, as described in my *Nature* paper from year 2001."[18] He added, "Recently we have discovered that water cleavage on mesoscopic iron oxide films proceeds at an amazingly high efficiency. This has given new impetus for this second pillar of our research. We have been able to obtain major European and Swiss research grants in this area and the number of our publications in this field is growing again."

As early as the early 1990s, NREL researchers began experimenting with a hydrogen-evolving photosynthetic bacterium, *Rhodospirillum rubrum*, that produces hydrogen via its hydrogenase enzyme from water even in the presence of oxygen. Normally, hydrogen-producing enzymes in water-splitting microbes are stopped dead in their tracks by oxygen, which inevitably evolves when the water molecule is broken up. But an NREL team led by the late Paul Weaver discovered a subset of *Rhodospirillum* that continues to pump out hydrogen even when exposed to atmospheric oxygen. Those organisms still do not split water, however. Coming up with a biological system that will split water entails a set of complex interlocking research tasks, including transferring the genetic properties of *Rhodospirillum* to another organism—a cyanobacterium, for instance. Beyond that, Weaver wanted to duplicate the processes that take place in a living cell in a cell-free system "where we take just the light-absorbing, water-splitting complex and link it directly to the hydrogen-producing enzyme," he said in a 1993 interview. In this particular approach, the theoretical efficiency of hydrogen production from water is quite low: about 10 to 12 percent at maximum. Real efficiencies in real cells are minuscule—only about 0.1 to 0.3 percent. But, said Weaver, microorganisms are cheap, and "cost is in favor of a biological system. Nature makes the catalysts for doing it very inexpensively from carbon dioxide and water, or from synthesis gas. Operating expenses would be equivalent to those of a farm producing grain crops."

Since then, enzyme-assisted hydrogen production R&D has come a long way. An experimental lower-temperature process to make hydrogen from biomass surfaced in early 2009. "Cocktail reception" was the snappy two-word opener for the abstract describing the new approach in the February 2 issue of the journal *Chemistry and Sustainability*. Researchers from laboratories at Virginia Polytechnic Institute and State University in Blacksburg, Virginia; the University of Georgia; and Oak Ridge National Laboratory described the process in which cellulosic materials from wood chips—cheap sugars or glucose will do nicely, but crop waste or switch grass would also work—are mixed with water and what they called a cocktail of a dozen-plus enzymes and then heated to a summer temperature (32°C) at normal pressures and swirled around in one pot. The product is hydrogen clean enough to run fuel cells. The obstacles at the time were the high costs of enzymes and slow reaction rates, but one of the paper's authors, Oak Ridge's Jonathan R. Mielenz, said in an e-mail exchange that lower-cost cloned enzymes are a possibility, and apparently there are other possibilities to improve the process.[19]

Another group of NREL researchers in Golden has been pursuing the path that Fujishima and Honda in Japan identified in the early 1970s. In early 1998, John Turner and Oscar Khasalev reported they had achieved a conversion efficiency of 12.4 percent by splitting water directly with sunlight using a combination of photovoltaic and photo-electro-chemical cells in a single step. In some test runs, "We have seen as much as 15–16 percent, but 12 percent is a good reproducible result," Turner said at the 1998 DoE Hydrogen Program Review. Their achievement, first announced in April 1998 in *Science*, made headlines all over the world after the Associated Press picked up the story.[20] *Chemical and Engineering News* quoted chemistry professor Nathan Lewis of the California Institute of Technology as saying that it was a "spectacular number."[21] Turner was more circumspect: it "isn't the magic bullet that gets us there," he said, but it is a "nice scientific accomplishment." Since then, things have gotten much more complex, and photoelectric water splitting with semiconductor materials is evidently more difficult than anticipated when it started in 1991 (NREL's project is scheduled to run through 2012). In his presentation at the 2008 DoE Hydrogen Program Review, Turner, by now recognized by DoE with the prestigious title "NREL research fellow," and five colleagues estimated that there are "easily 50,000 combinations of ternary oxides and almost 2 million quaternary oxides" that could be tried experimentally—a daunting task.

Summarizing the program's objectives a year later at the 2009 review, Turner and two colleagues said in their presentation that the purpose is to discover a semiconductor material or device that splits water into hydrogen and oxygen spontaneously upon illumination with a solar-to-hydrogen efficiency of at least 5 percent with a clear pathway to a 10 percent water-splitting system and "the possibility" of 1,000 hours of stability under solar conditions—and that it can be adapted to volume manufacturing. A year after that, at the 2010 review meeting, it still looked like a tough row to hoe: Turner, after being affectionately introduced by his session chairman as the "Grand Poobah" of photoelectrochemistry, called for, among other things, more work on theory —specifically, theoretical screening of oxides for photoelectrochemical hydrogen production. "We're convinced there is a material out there that will do that," said Turner, but "there is no satisfactory material yet. . . . We need new materials, new oxides." Observed Turner caustically, "Once upon a time combinatorial synthesis" for finding new promising materials "was very big. Let me make a very cruel statement: Combinatorial synthesis means you are clueless. . . . To go rambling through the periodic table is not a very productive thing unless you can focus on something specific. That's why I think theory is so important." He summarized his final slide: "A viable PEC [photoelectrochemical] water splitting system requires a unique material that satisfies several specific requirements" but "no known material is suitable." He acknowledged that "incremental progress has been made enhancing stability and efficiency," but "new materials must be synthesized and characterized." Concluded Turner, researchers are "not going to meet DoE technical targets with slight modifications of the usual [oxide] suspects."

5

Primary Energy: Using Solar and Other Power to Make Hydrogen

"The biggest eco-electricity initiative of all times."

That's how Germany's ZDF TV channel described the kickoff of a truly gigantic solar thermal electricity project in the summer of 2009 that aims at harvesting enough solar energy in North Africa's Sahara Desert to supply Germany, in addition to the rest of Europe, the Middle East, and Africa with electricity, shipped across the Mediterranean Sea with high-voltage undersea transmission lines. Desertec (www.desertec.org) started in 2007 as a concept linking the sunbelt of the Middle East and North Africa regions with the techbelt of Europe to fight climate change "in a way that is economically, technically and politically feasible." Developed by an offshoot of the Club of Rome group, the international Trans-Mediterranean Renewable Energy Cooperation network together with the German Aerospace Center DLR, the project matured to the point that one of the world's biggest reinsurance companies, Munich Re, called a meeting in July 2009 to get serious about what is expected to be a 400 billion euro (more than half a trillion dollars) undertaking.

As a first step, about a dozen major companies, including corporate giants such as Deutsche Bank, Siemens, and ABB and energy companies E.On and RWE, as well as representatives from the Arab League and Egypt's electricity and energy ministry, signed off on an initial three-year $2.5 million (1.8 million euros) planning and launch phase. Prince Hassan bin Talal of Jordan, a former president of the Club of Rome, said in a video message to the meeting that Desertec is essential for providing power for the region; otherwise 50 million Arabs could be out of work by 2050: "The partnerships that will be formed across the regions as a result of the Desertec project will open a new chapter in relations between the people of the European Union, West Asia and North Africa."[1]

Hydrogen will not be the energy carrier of choice, though, explained Desertec in its Web site FAQ: "In principle, hydrogen as an energy carrier has an advantage over electricity. However, the conversion of solar energy into hydrogen, and the re-conversion of hydrogen into electricity for the supply network would involve a loss of 50% of the original energy used. With HVDC [high-voltage direct current] transmission, only 10–15% of this energy is lost. There would also be pumping losses when transporting the hydrogen gas to Europe. Also, hydrogen would have to be generated from water which is rare in the desert. Therefore, it is more sensible to transport solar energy via HVDC transmission involving lower loss rates to Europe, where hydrogen can be generated."

Desertec is the latest, most spectacular solar electricity project so far. If it comes to fruition, it will be distinguished, aside from its size, by the fact that it is a commercial, everyday utility operation. The earliest U.S. solar power plant took shape three decades earlier: "First Large-Scale Solar Energy Project Nears Completion in US Desert," read a headline in the *International Herald Tribune* of November 15, 1977. The story told about progress on an experimental project, Solar One, that eventually was built near Barstow, California. Completed five years later, this central receiver pilot plant in the Mojave Desert bundled the combined reflective force of 1,818 sun-tracking mirrors, concentrating this solar energy on twenty-four receivers—solar water boilers—atop a 300-foot tower. These receivers turned water pumped up to them into 960°F steam, which was piped to a turbine generator, producing electricity. The steam was condensed back into water and sent back to the receivers. The principal purpose of this installation was to test the concept's basic validity and the prototype boiler module, a thermal storage system, and the control system. During the first two years of operation, the $144 million, 10-megawatt plant generated 7,700 megawatt-hours of electricity at a cost of about $150,000 per month. A retrofit that kept most of the main components in place was completed in early 1998. Renamed Solar Two, the plant continued testing runs for about two years until it was shut down in April 1999 because of decreased program funding.

On the other side of the Atlantic, similar but smaller projects were getting underway around the same time. In Sicily, the European Economic Community was completing a 1-megawatt solar electric tower directly connected to the local grid. In Spain, the Paris-based International Energy Agency was building two experimental solar energy plants, one a "solar chimney" plant in Manzanares, 150 kilometers south of Madrid, and the other a "solar farm" plant in which electricity was generated by an array of photovoltaic solar cells; the idea was to compare the efficiencies of the two concepts. The solar chimney was decommissioned in 1989 after eight years of operation, and although there have been proposals for others, none were ever built, as far as is known. Photovoltaic power plants have been built in many places, most of them in Europe. As of September 2010, the largest one, an 80-megawatt power plant, was in Sarnia, Canada, according to a Wikipedia site. Very big ones of up to 550 megawatts are being planned or under construction, most of them in the United States.[2] In general, solar power has been growing rapidly in recent years: as of spring 2009, 1.2 gigawatts of solar plants were under construction and another 13.9 gigawatts have been announced globally through 2014. Spain continues to lead the way, with twenty-two projects totaling 1,037 megawatts under construction, all of them scheduled for completion by 2010. In the United States, 5,600 megawatts of solar thermal power projects have been announced, according to a Wikipedia site in November 2010.[3]

Explaining the Solar-Hydrogen Link

The link between solar power in its various manifestations (wind, wave, and hydroelectric power; photovoltaic energy; and even energy from plant matter) and hydrogen energy is fuzzy in many people's minds. A common mistake, made even by some fairly knowledgeable people, is the belief that hydrogen itself is a source of energy. It is not. To be sure, hydrogen does exist in minuscule percentages in natural gas and therefore could, theoretically at least, be regarded as a "source," but that is useless in practical terms.

Hydrogen is, of course, an energy source in the nuclear sense. In the sun, hydrogen atoms fuse into helium to generate the energy that comes to Earth as solar radiation—nuclear fusion. But hydrogen is not a chemical energy source, although in the context of a hydrogen economy, it is thought of as a chemical fuel—a chemical analogue to electricity as an energy carrier with the added advantage that it can be stored over time and thus can help smooth out the fluctuations in the energy supply that are inevitably associated with intermittent primary sources such as the sun and wind.

Electricity is difficult to store in amounts large enough to meet the needs of a city, a region, or an industry (leaving aside techniques such as pumped storage—big reservoirs into which water is pumped when the demand for electricity is low so that it can be used to run turbine generators when the demand is high. Typically electricity has to be consumed the instant it is generated).

In addition, electrical energy accounts for only a small part of the world's total energy used, according to the U.S. Energy Information Administration's *International Energy Outlook 2009*. Among end users (residential, commercial, industrial, transportation), electricity accounted for only 55.2 quads (quadrillion Btu) out of a total of 343.6 quads delivered worldwide in 2006 (not counting 128.9 quads in electricity-related losses). Chemical energy in all its forms—liquids, natural gas, coal, and renewables—accounted for the much bigger share of 288.4 quads. And it does not look as if this ratio will change much in the future. For 2030, EIA projects that electricity will account for only 98.6 quads out of a total of 476.7 quads of delivered energy. This, then, makes it all the more urgent that these forms of chemical energy—almost all of them polluting in one way or another—be replaced by clean, nonpolluting hydrogen.

It is also useful to remember that usable electric power does not exist raw in nature but must be manufactured with complicated machinery drawing on primary sources (oil, coal, uranium, and, preferably now, solar radiation). *Usable* is the key word here; obviously there is electricity in nature, but it is hard to capture the power of a lightning bolt and make it run a piece of machinery. Similarly, hydrogen must be manufactured in a variety of ways—from natural gas (the lowest-cost, industrially preferred way), perhaps from coal, but eventually from water and variants of solar energy: photovoltaics, wind, direct solar, solar thermal, and, by extension, biomass.

We might compare hydrogen to the transmission in an automobile: it does not generate power by itself, but it makes it convenient to convert the power available from an engine into useful work. Hydrogen works in somewhat the same way but with some extra advantages.

Hydrogen also can also be conceived of as a broad river to which many primary source tributaries contribute.[4] At the downstream end, the broad hydrogen river splits again into multiple "irrigation canals" sustaining many economic activities, including transportation, various industries, domestic uses, and chemical activities such as making fertilizer and proteins.

Another useful way of looking at hydrogen may be to think of it as a man-made natural gas minus the global-warming-causing carbon. *Decarbonization*, a shorthand term that began cropping up among academic energy strategists in the early 1990s, describes the idea of industrially removing the carbon from fossil fuels before they are combusted and storing ("sequestering") the carbon in form of CO_2 in exhausted gas and oil fields, for example, before it can reach and pollute the atmosphere. Decarbonized fossil fuels are, of course, hydrogen. An important implication of this elegant and simple but difficult-to-implement idea is that various types of primary sources can be used for end uses that would not be possible without the hydrogen intermediary. Hydrogen can be used to convert nuclear energy, solar energy, or hydropower to chemical energy to propel cars and airplanes, for example. Without hydrogen, these primary sources are good only for making electricity; it is hard to imagine a large passenger plane powered by electric motors.

Storability is another important attribute of the chemical fuel hydrogen. Hydrogen makes it possible to economically store energy derived from intermittent sources such as solar power—in winter, for example. Hydrocarbons (natural gas, petroleum, coal) obviously are easy to store. But how do you store sunlight or the heat from a nuclear reactor? Storage works well in solar power tower plants, where heat is stored very efficiently in 24-hour, day-and-night cycles in molten salt storage tanks, but it does not work very well in retail applications, such as automobiles and trucks, or in long-term seasonal storage.[5] However, electricity can be stored indirectly if it is used to split water into hydrogen and oxygen, which can be stored easily like other industrial gases. By burning the hydrogen later with the stored oxygen or with oxygen from the atmosphere, electricity or heat, or both, can be produced again, when and where it is needed, by means of various types of combustion machines (internal combustion machines like a car engine or external combustion machines like a steam engine)—or, more efficiently, by recombining the two gases electrochemically, silently and cleanly, without any open flame, in fuel cells.

As far back as the 1970s, the eminent electrochemist John Bockris, founder of the Center for Electrochemistry and Hydrogen Research at Texas A&M University and in earlier years one of most active international proponents of a hydrogen economy, put the relationship between primary sources and hydrogen as follows:[6]

The likely sources of energy for the future are atomic and solar. Atomic reactors can provide electricity which would be cheaper as the reactors increase in size, but with size comes the difficulty of thermal pollution, so that large atomic reactors, which would give relatively cheap electricity at source, would have to be placed either on the ocean, far from population centers, or in remote areas such as Northern Canada, Siberia, or Central Australia.

Correspondingly, massive solar collectors are likely to be far from the population centers which need them, for they would be most advantageously situated in North Africa, Saudi Arabia, and Australia. Hence, the electricity to which they would give rise is liable to have to travel at least 1000 miles (1,600 kilometers), and, in some situations, as much as 4000 miles (6400 kilometers), to go from the site of production to the site of use.

The likelihood of this situation, and the energy loss in conduction, gave rise to the concept of a "Hydrogen Economy." Thus, it could be cheaper to convert electrical energy, which will be a product of solar and atomic reactors, to hydrogen at the energy source. Thereafter, the hydrogen would be transmitted through pipes—the pumping energy being relatively small—and converted back to electricity at the site of use (fuel cells) or used in combustion to provide mechanical power.[7]

Since the mid-1980s, nuclear power has lost much of its luster as a source of primary energy for electricity production.[8] Aside from continuing deep fears about nuclear power itself and its destructive potential and misgivings about storing highly toxic nuclear waste over tens of thousands of years, nuclear power has turned out to be a huge economic millstone around the collective necks of utility managers. Instead of "too cheap to meter" (the rallying cry of early nuclear enthusiasts), nuclear power has turned out to be "too costly to manage," Paul Gunter of the Washington-based Nuclear Information and Resource Service wrote in a compilation of articles by renewable energy specialists.[9] Gunter quoted a February 1985 article in *Forbes* magazine, "Nuclear Follies": "The failure of the US nuclear power program ranks as the largest managerial disaster in business history, a disaster on a monumental scale. The utility industry has already invested $215 billion in nuclear power, with an additional $140 billion to come before the decade is out, and only the blind, or the biased, can now think that most of the money has been well spent. It is a defeat for the US consumer and for the competitiveness of US industry, for the utilities that undertook the program and for the private enterprise system that made it possible."

In the new millennium, nuclear power continued to be promoted by the Bush and Obama administrations as a key weapon to fight climate change, but public reaction was still decidedly mixed. In November 2009, U.S. Senators Lamar Alexander (R, Tennessee) and Jim Webb (D, Virginia) introduced the bipartisan Clean Energy Act of 2009, a proposal to spend $20 billion over ten to twenty years to promote clean energies with a focus on nuclear power. "If addressing climate change and creating low-cost reliable energy are national imperatives, we shouldn't stop building nuclear plants and start subsidizing windmills," Alexander said in the joint announcement. "This legislation is a practical approach to move the United States toward providing clean, carbon-free sources of energy, to help invigorate the economy, and to strengthen our workforce with educational opportunities and high-paying jobs on U.S. soil," seconded Webb.[10] The *Los Angeles Times,* for one, was not impressed. Reacting to the bill and to other pronuclear efforts in Congress, a November 28, 2009, editorial called nuclear plants "a vastly expensive, inefficient and dangerous source of energy that requires massive taxpayer bailouts." Added the editorial, "Nuclear energy is not a reasonable solution because plants take too long to build and cost far too much. Actually, it's been so long since one has been built in this country that no engineering firm will even provide an estimate of the cost, but it's safe to say that each new plant would run to several billion dollars. Because lenders aren't willing to put up the money for such a risky investment, the nuclear industry is looking to Uncle Sugar." Nor were other critics impressed. Mark Cooper, a senior fellow for economic analysis at the Institute for Energy and the Environment at Vermont Law School, said in a June 2009 study that Wall Street was gun-shy and is declining to underwrite financially "risky and uneconomic" new nuclear reactors. In a November 2009 follow-up report, "All Risk, No Reward for Taxpayers and Ratepayers," Cooper, citing reports from Moody's and after analyzing what he says were more than three dozen cost estimates for proposed new nuclear reactors, concluded that Congress and state lawmakers would be well advised to follow the same course to avoid leaving taxpayers and ratepayers holding the bag. Coming nicely full circle, his study noted that "the last time the nuclear industry circumvented the judgment of the marketplace, it resulted in what *Forbes* magazine called the 'largest managerial failure in American history. The past could be prologue.'"[11]

However, there are serious environmentalists who argue that advanced nuclear power—and the key word is "advanced"—is well worth considering. James Hansen, perhaps the world's preeminent climatologist—he heads NASA's Goddard Institute for Space Studies and is Adjunct Professor of Earth and Environmental Sciences at Columbia University's Earth Institute—makes a case for clean nuclear power, with much lower chances of killing people than fossil fuels. In extensive footnotes in a 2011 paper on his website (http://www.columbia.edu/~jeh1/mailings/2011/20110729_BabyLauren.pdf) Hansen argues new Generation-4 technologies now under development (notably in China) are more fail-safe than the fifty-year-old Fukushima designs, and reprocessing nuclear waste (a resource really), offers clean energy for a few thousand years.

Internationally, the picture was mixed as well. A new-generation nuclear reactor going up in Olkiluoto, Finland, which had been described as the showpiece of a nuclear rebirth, was in difficulties. Constructed by a consortium of France's AREVA and Germany's Siemens, the 1,600-megawatt reactor was the most powerful reactor ever built; its modular design was supposed to make it less expensive and faster to build. The May 28, 2009, *New York Times* story, "In Finland, Nuclear Renaissance Runs into Trouble," said that "after four years of construction and thousands of defects and deficiencies, the reactor's 3 billion euro price tag, about $4.2 billion, has climbed at least 50 percent." Four months later, Germany's *Sueddeutsche Zeitung* newspaper, in its September 2 online edition, reported the "prestige object of the European atomic industry is sitting at the edge of the abyss": Siemens and AREVA threatened to halt construction of the project, behind at least three years and with large cost overruns. Still, interest in building new nuclear plants continues in places such as Abu Dhabi, and some new players have been coming onto the scene:

South Korea emerged as an unexpected contender in this field, winning a major order at the end of 2009 to build and operate four nuclear reactors with a total price tag of as much as $40 billion—the first nuclear plants in the Arab world.[12] Before the Korean team, led by Korea Electric Power Co. and including the construction units of Samsung and Hyundai groups, came into the picture with bids to build 1,400-megawatt reactors, only two groups had been in contention: a French group led by AREVA that had offered to build 1,600-megawatt reactors, and a Japanese-U.S. team consisting of General Electric and Hitachi with an offer to build 1,350-megawatt reactors.

And then there is nuclear fusion: a long-shot, dark-horse option but one that has been inching closer to reality. Unlike nuclear fission, fusion produces no radioactive fuel waste, although the reactor itself would become radioactive over time. Also unlike fission, fusion produces no bomb-grade materials, has no chance of inflating into a runaway critical reaction, and has a virtually unlimited fuel supply. Fusion reactors would run on seawater. The primary reaction requires deuterium and tritium, the two heavy isotopes of hydrogen. Deuterium is extracted from seawater, and tritium is made in a reactor from lithium, also found in seawater. Fusion would produce no local air emissions and no greenhouse gases—if scientists and engineers ever manage to develop an economic method of controlling the thermonuclear reaction. Hydrogen could be generated in a fusion reactor itself, providing another future pathway toward a sustainable energy future with no environmental damage and a virtually unlimited fuel supply.

In 2009, fusion was beginning to move glacially toward a real test with the start of construction of a huge 23,000-ton machine on a 104-acre platform, the so-called ITER (International Thermonuclear Experimental Reactor) in Cadarache, some 38 miles north of Marseille in France (the facility sprawls over 445 acres). The goal is to have the facility come to life with the generation of 150 million°C—ten times the temperature at the core of the sun— by 2018. But on the eve of the 2009 Copenhagen Climate summit, a slew of media reports about technical issues, timely manufacture of the huge components—each of ITER's eighteen magnetic confinement Toroidal Field coils that generate the plasma will weigh 360 tons, about the weight of a fully loaded Boeing 747–300 plane—cast doubt on that date.[13] Importantly, cost and money squabbles—the original budget of about 5 billion euros ($7.4 billion) is now expected to double—among the seven members (China, the

European Union, India, Japan, Korea, Russia, and the United States) indicated that the target is unlikely to be met. But if ITER works as its developers hope and expect, it would be a huge breakthrough: The target is a ten-fold net gain in energy, production of 500 megawatts of power with an input of 50 megawatts.[14]

For the moment, however, solar and renewables represent the best complementary energy resource for hydrogen production in terms of environmental benefits. Some of the more progressive fossil fuel companies began seeing the light in the 1990s. John Browne, at the time chief executive officer of British Petroleum, the world's third-largest oil company, was one of the first among his peers to sign on as a believer in the reality of CO_2-caused global warming and setting his company on a renewable-energy course, according to an article in *Business Week*, "When Green Begets Green," on November 10, 1997. Browne believed renewables could meet 5 percent of the world's energy demands in twenty years and 50 percent by 2050 and that BP should get in on the ground floor. A key component of its alternative energy business has been the San Francisco–based BP Solar subsidiary, helped by its catchy "Beyond Petroleum" slogan. However, media reports said BP Solar shut down several solar plants in the United States and Spain in 2009 and has shuttered its alternative energy headquarters in London.[15] Nevertheless, it is keeping more than a hand in hydrogen—specifically, coal to hydrogen. Together with its Australian coal-producing partner Rio Tinto, it set up a new jointly owned company, Hydrogen Energy, to build a natural gas-to-hydrogen power plant in Abu Dhabi. A second project that surfaced in 2009 is the Hydrogen Energy California Project, a 390 megawatt integrated gasification combined cycle power plant, including carbon capture and storage, that BP wants to build in Kern County, California.[16] The Royal Dutch/Shell Group also broke ranks with fossil fuel orthodoxy. In quick succession, it set up a new $500 million Renewables division in autumn 1997, followed in 1998 by a decision to quit the exquisitely misnamed Global Climate Coalition (an association of oil and other companies dedicated to the proposition that global warming has no relation to increasing fossil-fueled CO_2 emissions). Shortly after that, Shell began to take a serious look at hydrogen. In June it dispatched a platoon of executives to the 1998 World Hydrogen Energy Conference in Buenos Aires for a behind-the-scenes briefing, followed in quick succession by a decision to set up a separate Shell Hydrogen division. Over the years, Shell Hydrogen has been active in setting up hydrogen fueling stations in various parts of the world and in other activities, although it slowed the pace beginning in 2009 with the global financial crisis, including elimination of the Hydrogen division as a separate entity during a corporate reorganization.[17]

In principle, solar energy can be employed to make hydrogen in a variety of ways. Typically electricity is produced first; it then can be used in electrolysis. It is not clear whether there is one best solar technology for making hydrogen, and many variables come into play, including capital costs, the commercialization status of a given technology, local circumstances (e.g., special tax breaks), the total chain of conversion efficiencies from the amount of sunlight available in a given location, average wind speeds, the growth rate of plants (for biomass production), the conversion efficiencies of various types of photovoltaic cells, the efficiencies of water splitting and hydrogen storage, and the price and usefulness of the by-product oxygen.

A large body of information relating the various primary sources to hydrogen production is available in a series of papers and studies on hydrogen energy systems published since 1989 by the Center for Energy and Environmental Studies at Princeton University, many of them under contract to the U.S. Department of Energy (DoE). Many of those reports listed Joan Ogden as principal author or coauthor. Ogden started as a nuclear fusion physicist; however, after recognizing the difficulty of the scientific and engineering challenges facing fusion, she decided that she wanted to better understand energy and environment problems generally and "all the options for a clean energy future." She abandoned fusion and began working on hydrogen in 1985. Since then, she has become arguably the world's premier systems analyst for hydrogen energy, advising and consulting for the U.S. DoE and the State of California, the European Union, foundations, and automotive and energy companies, among others. She joined the University of California Davis Institute for Transportation Studies in 2003, where she is codirector of the Sustainable Transportation Energy Pathways program and a professor of environmental science and policy.

One of Ogden's early 1997 papers looked at the technical feasibility and the economics of a reasonably realistic hydrogen refueling infrastructure for at least a portion of the zero-emission vehicles that were expected to be mandated in southern California. Ogden's estimates of energy requirements were predicated on the higher efficiencies of fuel cells, not on the lower efficiencies of conventional internal combustion cars running on hydrogen. For starters, she said at the time, potentially ample sources of hydrogen were already in place in the Los Angeles Basin: hydrogen plants and refineries.

By 2010, the South Coast Air Basin was predicted to be home to some 9 million cars and 11 million counting sport utility vehicles (the actual number in 2007, the latest available, was 8.7 million light-duty vehicles).[18] Of these, more than 700,000 were expected to have to qualify as zero-emission vehicles, she wrote, if one assumes an "aggressive commercialization scenario" for fuel cell vehicles beginning around 2003 (which, of course, has not happened; only a few dozen fuel cell vehicles were operating in California in 2010). Several hydrogen plants in the area are producing large quantities of hydrogen. Although most of their output is already committed to customers, there is some extra capacity. Hydrogen produced with that extra capacity could be sold as transportation fuel. Also, a number of oil refineries in the area are producing huge amounts of hydrogen; most of it is used internally in refining, but several million cubic feet typically are available for off-site sale. With new hydrogen facilities planned to meet the expected demand for more reformulated gasoline, extra reformer capacity is being constructed now. Thus, said Ogden, hydrogen from existing sources "could be significant in getting hydrogen vehicles started." In addition, she wrote, "ample natural gas resources" are available in the Los Angeles area. And, she said, southern California has a "large potential" for using off-peak power to electrolyze water. That last assumption, however, is probably no longer totally valid in southern California because of growing concerns about issues connected with charging large fleets of plug-in hybrids and battery cars at night.[19]

Ogden and her colleagues have recently focused much more on strategies for managing phasing hydrogen and fuel cells into road transportation. One aspect of her work has been to compare the pros and cons of hydrogen and fuel cell vehicles versus plug-in hybrids and pure battery electric vehicles, with the latter getting much more public attention and support from the government. This is a contentious issue heatedly debated by fiercely opposing camps at the end of decade in both the United States and Europe. The plug-in hybrid/battery camp argues that these technologies are ready for use in the United States now and at commercially acceptable cost; the hydrogen/fuel cell folks counter that the world will never get to a sustainable level of low CO_2 emissions without these technologies.

Ogden examined these issues in a paper at the 2009 National Hydrogen Association annual meeting in Columbia, South Carolina. Among her topics were the expected major role of hydrogen and fuel cells past 2025; expected improvements in internal combustion engine technology; the role of biofuels and of plug-in hybrids past 2025; phased introduction of hydrogen fuel cell vehicles in American "Lighthouse" cities (a term used to denote cities where new advanced energy or environmental technologies are first introduced); financial assumptions; and hydrogen transition modeling. She asked, "What are investment costs for H2 fuel cell vehicles to reach competitiveness with gasoline vehicles? Cash flow analysis to see when strategy of introducing H2 fuel cell vehicles breaks even with Business As Usual (BAU)?" Among her conclusions were that "long term GHG [greenhouse gas] and oil use reductions are significantly greater with FCVs [fuel cell vehicles] than PHEVs [plug-in hybrid electric vehicles] for similar levels of energy supply decarbonization; transition costs, timing to 'breakeven year' are similar for FCVs and PHEVs (10s of billions of dollars total, spent over 10–15 year period); the majority of transition costs is for vehicle buydown (80%); infrastructure costs are not zero for PHEVs ((\$800–2,000/car for residential charging)." Summarized Ogden, "Both batteries and fuel cells are in the same ballpark. If they tell you it costs millions of dollars for fuel cells, and batteries are free, don't believe them."

Another paper Ogden coauthored for the University of California Davis Institute of Transportation Studies in late 2009 summarized the findings of five workshops held over fifteen months that outlined a road map and transition strategy for hydrogen and fuel cell vehicles in California through 2017. There would be participation or sponsorship from carmakers General Motors, Toyota, Daimler, and Honda; energy companies Shell Hydrogen and Chevron; and officials from the California Air Resources Board, California Energy Commission, South Coast Air Quality Management District, and National Renewable Energy Laboratory.[20] Noting that the coming 2050 targets for deep carbon reductions on the order of 80 percent will require a significant market share of zero-emission vehicles, the paper says there are only three energy carriers that can achieve that: advanced biofuels, hydrogen and electricity: "Advanced biofuels will play a significant role, but there will be limits on both physical supply and the use in light-duty vehicles compared to other sectors," the executive summary noted. "If H2 is not part of that fuel mix because of lack of early investment, the only significant option would be electricity used in battery vehicles. Relying on these limited options increases the risk of not achieving our long term climate and transportation energy goals, given current technical and market uncertainties." This will require major investments, and that's problematical: "The automotive and energy industries have the capabilities to make these investments, but their commitment to this alternative is fragile for two reasons. First, political momentum for hydrogen has declined at the time when larger public and private cooperation is needed. Second, introducing hydrogen and fuel cell vehicles will require a distinct coordination between new H2 stations and planned vehicle placements."

All this is iffy, the forty-eight-page paper noted. No true long-term carbon policies have been set at the time of writing (fall 2010), making it difficult to develop a viable business case. Said the executive summary, "Currently, a national level commitment to hydrogen and fuel cell vehicle technologies is uncertain. This is evident by the U.S. DOE's May 7, 2009 budget request, which virtually eliminated funding for R&D on hydrogen storage, production and delivery and for fuel cell vehicle demonstrations. Although Congress restored the funding, the initial cuts increase the risk to the industry's hydrogen and fuel cell investments. . . . Automotive FCV investments need to be supported now through timely early infrastructure deployment, to ensure production vehicle programs are launched soon enough to prepare for sales in 2015. Given the uncertainty of the U.S. DOE's support for hydrogen vehicles, state initiatives take on additional importance."

Still, UC Davis researchers have developed a ten-year launch and growth scenario for the Los Angeles Basin through 2017 that included an analysis of network convenience, station cost and infrastructure, and cash flow. It concluded that for an infrastructure of forty-two (or fifty-five) hydrogen stations in the region supporting 25,000 (40,000) fuel cell vehicles by 2017, total investments of $200 million ($260 million) would be needed. Amortized over ten years, an average hydrogen fuel retail price of $10 per kilogram could recoup the cost, although California's renewable hydrogen production requirement would impose additional costs.

There is an environmental downside to electrolytic hydrogen if it is produced with commercial electric power. Sandy Thomas, the founder of H2Gen Innovations, estimates that greenhouse gases in the United States would more than double if run-of-the-mill electricity were used to generate hydrogen, even if that hydrogen were to be consumed in efficient fuel cell vehicles. The reason is that more than half of the electricity in the United States is produced by burning coal. Using more coal-derived electricity to make fuel for cars and trucks would make a bad situation worse. "Hydrogen is not always a net benefit, and in this case can exacerbate climate change," according to Thomas. Even in California, which has proportionately fewer coal sources imported from outside the state than the rest of the country, hydrogen consumed in a fuel cell vehicle "would generate nearly the same greenhouse gases as conventional gasoline cars," Thomas said in a 2009 personal communication.

Primary Renewable Sources

Hydropower

Energy extracted from falling water is still the leading renewable energy source in the United States, according to an overview paper by Linda Church Ciocci, executive director of the Hydro Research Foundation.[21] In the past decade, however, its share has been declining. In the 1990s, it accounted for about 10 percent (about 90,000 megawatts) of the electric power generated in the United States and almost 98 percent of the renewable energy. Although the total was still about the same at the end of 2009, the relative numbers had dropped to about 6 to 8 percent and 75 percent, respectively, presumably because of faster growth of other energy types, including renewables. Still, the National Hydropower Association sees good growth prospects in the decades ahead: a potential capacity increase of 23,000 megawatts by 2025 in conventional technologies and another 13,000 megawatts from new technologies such as ocean and wave energy devices and new hydrokinetic technologies under business-as-usual assumptions. There is potential for as much as 60 percent more in that period, given the right policies supporting R&D, long-term investment, a federal renewable energy standard, and recognition of hydropower and other renewables under national climate policies, according to an association spokeswoman.

Ironically, hydropower is growing more slowly than other forms of electricity production in the United States, according to the Hydro Research Foundation. Its share will fall to less than 7 percent by 2015. But in the world as a whole, interest in hydropower is rising. It already accounts for 19 percent of the world's net electricity generation and 24 percent of the world's capacity. Some 131,000 megawatts of hydroelectric capacity are planned for Central and South America and another 127,000 megawatts for Asia. The biggest hydroelectric plant today (12,600 megawatts) is at Itaipu, on the Parana River in Brazil.

In terms of hydrogen production, an ambitious hydroelectric project, which may become significant for a future hydrogen economy, is proceeding in Iceland, where a parliamentarian named Hjalmar Arnason assembled a task force in the 1990s that essentially proposed converting the country's cars, trucks, and fishing fleet to fuel cells and hydrogen. Iceland's economically exploitable hydroelectric reserve is estimated at about 30,000 gigawatt-hours per year, of which only a small share has been utilized.[22] In1998, Iceland, Daimler-Benz, and Ballard began laying plans for a hydrogen economy, and in early 1999 the Icelandic Hydrogen and Fuel Cell Company was formed, with three multinationals—DaimlerChrysler, Royal Dutch/Shell Group, and Norsk Hydro—as minority partners.[23] The project continues despite some setbacks, caused in part by the global financial crisis of 2008 and 2009.

Wind

Next to hydropower, wind energy is generally regarded as the most cost-effective, least capital-intensive form of alternative primary energy, and both wind and solar have chalked up steeply accelerating growth in the last couple of decades. According to the Global Wind Energy Council (GWEC) Web site, global wind power capacity has grown at an average cumulative rate of more than 30 percent, with 2008 registering a record year with more than 27 gigawatts of new installations for a worldwide total of more than 120 gigawatts compared to 6.1 gigawatts in 1996.

Germany, which had been the world wind power champion, was relegated to second place that year with a total installed capacity of 23.9 gigawatts compared to the new leader, the United States, with a total of 25.1 gigawatts. That year American firms set a record by installing 8,500 gigawatts in new capacity, an order of magnitude more than the mid-1990s, when total installed capacity was about 1.75 gigawatts (1,750 megawatts) . In Asia, China doubled its total wind power capacity for the fourth year in a row, for a total of more than 12 gigawatts, according to GWEC; India's installed capacity was 9.6 gigawatts.

The global credit crisis and economic downturn that began in 2008 put a crimp in the rosy growth picture, however. In early 2009, the *New York Times* said that except in isolated cases such as China, "installation of wind and solar power is slowing, and in some cases plummeting, with trade groups projecting 30 to 50 percent declines."[24] The American Wind Energy Association (AWEA) predicted at least 5,000 megawatts would be commissioned in 2009, with the recession's impact cushioned by a renewable energy production tax credit under the 2009 American Recovery and Reinvestment Act. Internationally, the GWEC's 2008 report includes a five-year forecast predicting that wind energy capacity will triple by 2013 to 332 gigawatts.

Direct-Solar Electricity Generation

One of the most successful early demonstrations of harvesting the sun's power to supply electricity to homes and industry involves nine solar stations in the Mojave Desert, between Los Angeles and Las Vegas, at Daggett, Kramer Junction, and Harper Lake, California. These plants, producing 354 megawatts, used parabolic trough mirrors—the same basic technology now pursued by the huge Desertec concept but different from the Solar One and Two solar towers mentioned already in this chapter. The first went on line in 1984; construction of the last one ended in 1990. Designed and built by the American-Israeli firm Luz International, these plants have been converting solar power to electricity at peak efficiencies of about 21 percent, with an annual average efficiency of 12 percent and current costs of 8 to 10 cents per kilowatt-hour—not cheap enough for hydrogen production.[25] Recent data have been more promising. A 2010 draft case study by Argonne National Laboratory for producing hydrogen from solar energy using a copper-chlorine process in a 100 megawatt electric (Mwe) solar power tower predicted a cost of 6.8 cents per kilowatt-hour, based on heliostats costing $127 per square meter, the key parameter. A 2007 Sandia National Laboratories heliostat cost-reduction study, assuming 2006 labor and materials costs for a power tower of about 600 megawatts and heliostat costs of $126 per square meter, came up with electricity costs of 6.7 cents per kilowatt-hour and hydrogen costs of $3.20 per kilogram; further R&D should lead to heliostat costs of $90 per square meter, translating into 5.6 cents per kilowatt-hour and $2.75 per kilogram of hydrogen, this analysis said. The Desertec Web site says that the German Aerospace Center DLR, which is deeply engaged in Earth-bound alternative energy technologies, has calculated that if solar thermal power plants were to be constructed in large numbers in the coming decades, the estimated cost would come down to about 4 to 5 eurocents per kilowatt-hour (5.48 to 6.85 cents in early 2010).

Photovoltaics

Spurred by the oil shocks of the 1970s, the U.S. government has been supporting the development of photovoltaics ever since. As a result, the prices of photovoltaic modules have dropped from about $100 per peak watt four decades ago to just over $4 (cost of $3 per peak watt) at the end of 2008, but up from an average of about $3.65 a year earlier, according to a January 2009 report by Deutsche Bank Research, "Solar Photovoltaic Industry—(PV) Looking through the Storm."[26] Due to growing demand for renewable energy sources, PV production has doubled every two years between 2002 and 2008, making it the world's fastest growing energy technology in those years according to a periodically updated Wikipedia site.[27] In 2008, cumulative global PV installations had reached 15.2 gigawatts, with about 90 percent consisting of grid-tied systems, according to the entry (by 2009, grid-connected PV had reached some 21 gigawatts, a 2010 update said). But with the onset of the global economic meltdown, Deutsche Bank analysts were expecting a "brutal" shakeout because of oversupply and acute credit tightness for at least early 2009: tellingly, the analysts scaled back their year-to-year demand growth expectations (megawatt peak installed) to be only about 10 percent well below their mid-2008 forecast of 38 percent.

Solar cell efficiencies have increased dramatically from about 4 percent in the mid-1970s to a market average of 12 to 18 percent and as much as 23.4 percent for a company widely regarded as market leader, SunPower, based in San Jose, California. Efficiencies of more than 40 percent have been achieved with concentrated light PV systems at the University of Delaware in a joint project with DuPont;[28] Sharp Corp. announced 35.8 percent efficiency without concentration in 2009, using a proprietary triple-junction manufacturing technology.[29]

Whether PV can produce electricity at costs low enough to generate hydrogen economically as transportation fuel looks doubtful; it is not really the goal of PV companies anyway. One recent survey by a solar consultancy, Solarbuzz, quoted June 2011 prices from a high of 66.68 cents per kilowatt hour (residential system, cloudy climate) to a low of 16.11 cents (industrial system, sunny climate).[30]

Biomass

In the past decade, biomass has gained a lot of public attention as a new, environmentally benign source of liquid fuels for transportation—cars, trucks, buses, and also, phasing in very slowly, air and rail transport. But on the global scale, these uses are still trailing way behind the traditional main use of providing heat. A 2007 report by the MIT Joint Program on the Science and Policy of Global Change, "Biomass Energy and Competition for Land," reported that biomass provided 39 exa joules (EJ; 1 EJ corresponds to 24 million metric tons of oil equivalent, [Mtoe]) for cooking and heating, or 9.3 percent of global primary energy use; electricity and fuel generation was estimated to account for only 6 EJ, a mere 1.4 percent. A 2009 International Energy Agency (IEA) Bioenergy report, "Bioenergy—A Sustainable and Reliable Energy Source," puts the world total a bit higher, about 50 EJ, or 10 percent of the world's total primary energy. Estimates of the world's bioenergy production potential vary widely, from a low of 350 EJ per year to 2,900 EJ, according to the MIT report.

The IEA paper says that forestry, agricultural, and municipal residues and wastes are the main feedstocks for the generation of electricity and heat from biomass, adding, "A very small share of sugar, grain, and vegetable oil crops are used as feedstock for the production of liquid biofuels," such as biodiesel. In the medium term, lignocellulosic crops such as switchgrass, hybrid poplar and willow, could be produced on marginal or surplus land. In the longer term, aquatic biomass—algae— could make a significant contribution, according to the IEA paper. Also, gases generated in landfills from municipal solid wastes can be burned with coal in conventional coal-fired power plants, burned separately in steam plants, and run gas turbines, internal combustion engines, and fuel cells. Companies such as FuelCell Energy, based in Danbury, Connecticut, have successfully demonstrated fuel cell power plants running on landfill gas.[31]

Biomass-fired power plants typically use small steam turbines (10 to 30 megawatts; some go as high as 50 megawatts), because usually the amount of biomass at a given site is relatively small. Since these plants are small, they cannot economically incorporate sophisticated heat-recovery components, and their efficiency is therefore relatively low: 17 to 23 percent, with some larger ones achieving 28 percent. (A large modern coal plant gets up to about 35 percent.) Still, direct-combustion biomass plants have produced electricity at fairly low cost of about $0.073 per kilowatt-hour.[32]

Higher efficiencies can be achieved by first gasifying biomass and then running it through a high-efficiency gas turbine system. Systems that do this (now under development, primarily for coal) are expected to be quite cost-effective, making electricity at somewhat lower costs—about 7.1 cents per kilowatt-hour. By 2020, electricity from gasified biomass is expected to cost only about 4.3 cents per kilowatt-hour.

There are projects to demonstrate the use of marginal or underused lands to raise fast-growing energy crops in "crops-to-power." One of them, in western Minnesota, burns alfalfa stems (about 50 percent of the crop); the leaves are used for animal feed. A second involves growing so-called switchgrass on 40,000 acres of marginal cropland in Iowa to provide 35 megawatts of co-fired power in a 735-megawatt power plant. (Switchgrass is a deep-rooted indigenous prairie grass that helps to build up and sta-bilize the soil; it efficiently reseeds itself and does not need pesticides.)

A third, initiated by what was then Niagara Mohawk Power Corpora-tion, involves the cultivation of shrub willow trees as a carbon-neutral renewable feedstock for combined heat-and-power (CHP) plants as well as a fuel supplement to coal-fired power production. A consortium is growing willows on plantations located in a 25- to 50-mile radius from selected CHP and coal power plants in upstate New York. Willows can reach heights of 28 feet in three years, after which they can be harvested with no need for reseeding; the consortium expects seven harvests in twenty-one years from one planting of improved willow cuttings from a commercial willow nursery in Fredonia, in western New York, Double A Willow, which operates more than 100 acres of nursery beds for commercial production of willow biomass crops. A willow bioenergy crop breeding program is underway at Cornell University's New York State agricultural experiment station in Geneva, New York, and agricultural equipment maker Case New Holland is developing a specialized cutting head for willow crops that fits on their forage harvesters so that the crop can be cut and chipped in a single pass.

Biomass does produce emissions, including nitrogen oxides, particulates, volatile organic compounds, and various toxics. It has been argued, however, that these pollutants are well understood and treatable and that emissions should meet air quality standards.[33]

But the major attraction of biomass energy, from the perspective of reducing greenhouse gases, is that it is carbon neutral: the CO_2 generated when biomass is burned will be captured again with new plant growth—an elegant match with the natural carbon cycle. Critics have pointed out, though, that net carbon emissions do result if the plants are not regrown. Also, farm equipment, trucks, and other machinery needed to harvest, transport, and process biomass all use fuels. If those fuels are gasoline or diesel, carbon emissions result. If they are biomass derived, they cut overall biomass energy production.

There is a debate as to whether and how much biomass could contribute to the world's energy supply. As a key message, the IEA paper's executive summary says that bioenergy "could sustainably contribute between a quarter and a third of global primary energy supply in 2050. It is the only renewable source that can replace fossil fuels in all energy markets—in the production of heat, electricity, and fuels for transport." The older MIT study was more circumspect and skeptical: "Because global demand for food is also expected to double over the next 50 years, increased biofuel production competes with agricultural land needed for food production."[34] It added, "The idea that biomass energy represents a significant domestic energy resource in the USA is misplaced. If the USA were to actually produce a substantial amount of biofuels domestically, through policies that spurred its use but that prevented imports, instead of relying on oil imports, the country would need to rely on food imports." Overall, it concluded, "The scale of energy use in the USA and the world relative to biomass potential is so large that a biofuel industry that was supplying a substantial share of liquid fuel demand would have very significant effects on land use and agricultural markets."

Advanced Solar Concepts

Coming up slowly, behind more conventional technologies, are new ones that aim to employ solar power through novel (and, it is hoped, more efficient) ways to make electricity or hydrogen. One of them is direct high-temperature solar water splitting, literally the hottest and technically one of the most difficult methods. The basic idea is to use highly concentrated sunlight, focused by a large number of ground-level mirrors onto a point atop a solar tower, to generate temperatures of thousands of degrees, at which steam dissociates into hydrogen and oxygen.

Attempts in France, Canada, and Japan in the late 1970s and the early 1980s to split water directly using focused sunlight typically converted only about 10 percent of the steam into its constituent gases. A team at the Weizmann Institute of Science in Israel, headed by Abraham Kogan, tackled the problem again in the early 1990s, believing it could do much better and perhaps even develop an industrially viable process. In a 1996 interview, Kogan explained that previous efforts had attempted direct thermal water splitting at 2,500°K (2,227°C) and normal atmospheric pressure.[35] "We intend to operate at 2500°K and 0.05 atmospheres, a state at which 25 percent of the water is split," he said, but apparently this was never achieved. The work nevertheless continues at the Weizmann Institute under new leadership, notably Jacob Karni, director of the Center for Energy Research. It looks as if high-temperature direct solar water splitting has run into a brick wall, but work is continuing on solar thermochemical water splitting (see chapter 4), and also high-temperature solar dissociation of methane, both under investigation at the National Renewable Energy Laboratory in Golden, Colorado.

Uncommon Sources

Since the 1970s, in the aftermath of the early oil shocks and again in the wake of climbing energy costs that began in the 1990s and sharply increased after 2000, various imaginative, perhaps esoteric, and even outlandish alternative energy schemes have cropped up. Some examples follow.

Giant Solar Chimneys

Large-scale production of electricity from solar power is the goal of the solar chimney idea proposed by two West German engineers, Jörg Schlaich and Rudolf Bergermann, as early as 1976. The basic idea is simple: it combines the greenhouse effect (air and soil are heated underneath the translucent collector roof by solar radiation) with the chimney effect (a strong upward air draft created by the temperature differential drives wind turbines and power generators). Such plants, claimed Schlaich, could produce electricity at competitive costs.

Schlaich and Bergermann envisioned a huge hybrid industrial smokestack and cooling tower, up to 3,300 feet high and 330 feet or more in diameter. The chimney would be built of lightweight concrete or would be assembled from cable mesh and aluminum cladding hung from a central concrete pillar. At ground level, the chimney would be surrounded by a translucent roof of glass or plastic, which would be 6 to 18 feet above the ground and up to about 7 miles in diameter. As an added benefit, the roof's peripheral areas, where the air mass drawn to the center would be only a gentle breeze, could be used to grow crops.

A first experimental solar chimney plant was completed in 1983 near Manzanares, about 100 miles south of Madrid. It ran successfully until 1990. In 1996, planning started for a considerably larger plant (200 megawatts) in the Indian state of Rajasthan, but it never got to construction. A drawback of solar chimneys is that their efficiency in converting solar energy to electricity energy is fairly poor. For the 200-megawatt plant planned for India, Schlaich's office projected an efficiency of only about 3 percent—a lot less than the double-digit efficiencies claimed for the trough solar plants in the Mojave Desert.

In 2002, EnviroMission, a company in Armadale, Victoria, announced plans for four kilometer-high solar chimneys, each producing 200 megawatts. The Australian government granted it major project facilitation status, a designation that was supposed to give it more credibility with investors, but the plans never came to fruition. Since then the company has shifted its attention to the American Southwest, announcing in 2007 plans for building smaller chimney towers in Arizona.[36] The project appears to be moving forward. The October 28, 2010, *Phoenix Business Journal* reported that EnviroMission signed a deal with the Southern California Public Power Authority, a utility group representing eleven municipal utilities and one irrigation district in southern California, to buy electricity. The electricity will be generated by a 2,400-foot, 200-megawatt solar tower the company plans to build in western Arizona by 2014 pending approval by the Arizona Corporation Commission, along with meeting local permitting requirements and securing financing.

Orbiting Solar Power Stations

An idea that has received some media attention over the decades is to collect solar energy by means of orbiting solar power stations, which would then transmit huge quantities of electric power to Earth in microwave beams. In an alternative scheme, an orbiting system of mirrors would reflect and focus sunlight onto Earth-based solar receiver power plants, which would convert the solar energy into electricity.

Boeing Aerospace and Rockwell International were said to be exploring the first idea in the 1970s. NASA and the DoE spent some money on a multiyear feasibility study. A 1978 article in the magazine *Engineering News-Record* quoted Peter Glaser, then president for engineering at Arthur D. Little, as saying that such systems could supply as much as 25 percent of the planet's energy requirements by 2025.[37]

The second approach—putting a number of mirrors in orbit to beam sunlight onto an "energy oasis" about 25 miles in diameter—was outlined in 1977 by Krafft Ehricke, a former NASA manager.[38] Ehricke said that "an industrial sun for Europe" would create a climate in the "energy oasis" roughly equivalent to that of the Arab or the Australian desert, and that the total yield from the "oasis" would be between 35,000 and 50,000 megawatts. The idea never caught on.

Now it appears that in a few years, solar power may in fact flow down to earth from space. The *New York Times* "Green Inc." blog reported on December 3, 2009, that the California Public Utilities Commission has approved a utility contract for the country's first space-based solar plant. The partners are California's big Pacific Gas and Electric (PG&E) utility and a Manhattan Beach, California, start-up, Solaren Corp., founded by space veterans from Hughes Aircraft, Boeing, and Lockheed. The basic idea is to deploy a huge inflatable Mylar mirror, 1 kilometer in diameter, which will collect and beam sunlight onto a smaller mirror that in turn would focus solar radiation onto terrestrial PV modules at a ground station, which the *Times* account said would be located near Fresno. Total capacity is described as 200 megawatts, and the electricity would be sold at rates that so far are confidential to PG&E under a fifteen-year contract, with power deliveries to start in 2016.

PG&E and Solaren are not alone in pursuing space electricity: The utility's "Next100" blog reported on February 2, 2010, that Japan's Mitsubishi and more than a dozen other Japanese companies are working on it for the Japan Aerospace Exploration Agency. And Eads Astrium, described on the blog as Europe's number one space company, has said it has also started developing key components to beam power collected by orbiting solar panels back to Earth.

Converting Thermal Energy from Oceans

Ocean thermal energy conversion (OTEC), a technology that exploits the temperature differences between warm near-surface layers of the oceans and deeper, colder layers, never took off despite intermittent high hopes over the decades. A French physicist, Jacques d'Arsonval, first suggested the basic idea more than 100 years ago. In the early 1930s, a French engineer, Georges Claude, built an ocean-thermal plant on the shore of Matanzas Bay, about 50 miles west of Havana, that produced about 22 kilowatts of electricity. Claude also built a plant aboard a cargo vessel moored off the coast of Brazil. Both plants were eventually destroyed by weather and waves, and neither appears to have produced more electricity than was needed to operate them. Cheap fossil fuels contributed to their demise.

About twenty-five years later, French researchers designed a 3-megawatt OTEC plant for Abidjan (on Africa's west coast), but the plant was never completed. In 1979, the Natural Energy Laboratory of Hawaii (at Keahole Point, on the Kona coast of the island of Hawaii) built a 50-kilowatt plant on a U.S. Navy barge a short distance offshore. The plant produced 52 kilowatts of electric power, but once again, most of the power was needed to run the system—the net output was only 15 kilowatts, according to the National Renewable Energy Laboratory's OTEC home page. Another OTEC plant was built in 1981 by Japan on the island of Nauru. That 100-kilowatt plant generated about 32 kilowatts net power. And as recently as 1993, an "open-cycle" plant at Keahole Point produced 50 kilowatts of electricity in an experiment.[39]

Except for the facilities in Hawaii, OTEC had almost faded away, reported Luis Vega, a Hawaii-based OTEC researcher, in 1997. "Presently there is no OTEC activity in the world," he said.[40] A decade and a half later, there have been advances in technology details such as better ideas for heat exchangers, Vega said in an e-mail communication, but not much else, despite its huge potential: "I am extremely positive about OTEC but unhappy with unrealistic claims." In a 2002 paper that he described as still valid in 2009, Vega wrote that the amount of solar energy absorbed by the oceans is estimated to be equivalent to at least 4,000 times the amount of energy currently consumed by humans. With OTEC efficiency of only 3 percent, "we would need less than 1 per cent of this renewable energy to satisfy all of our desires for energy," he wrote, but no commercial-size OTEC plant has been built.[41]

Still, there is some progress, in part because of the 2008 spike in oil prices and because of Navy interest in developing dependable power supplies—unlike intermittent sources like wind and solar, OTEC would run all the time—for island bases in places such as Hawaii and Diego Garcia. That year, Hawaii's governor, Linda Lingle, announced a partnership joining Hawaii, Lockheed Martin, and the Taiwan Industrial Technology Research Institute to build an OTEC pilot plant in Hawaii off Kahe Point on Oahu. Lockheed's director of alternative energy programs development, Ted Johnson, said in a September 2, 2009, presentation at the Asia-Pacific Clean Energy Summit in Honolulu that this 10-megawatt pilot plant could be up and running in four years and could be scaled up to 100 megawatts in two more years, provided the pilot plant works out well. And in Washington, DoE's deputy assistant secretary for renewable energy, Jacques Beaudry-Losique, told the House Subcommittee on Energy and Environment at a December 3, 2009, hearing, the department had awarded about $2.6 million in fiscal years 2008 and 2009 to four projects focused on OTEC.

Geothermal Energy

Geothermal energy is the real sleeper—and a giant one at that—among all the sustainable energy source options. Geothermal electricity has been produced continuously since 1904 in the venerable Larderello geothermal fields in Italy's Tuscany region, some 40 miles southwest of Florence, according to a 1998 brochure, "Geothermal Energy," issued by the University of Utah's Energy and Geoscience Institute. Another 1998 publication, this one produced by the DoE's Office of Geothermal Technologies, says that geothermal electricity has been generated without interruption in New Zealand since 1958 and at The Geysers field in California since 1960. "In fact, no geothermal field has been abandoned because of resource decline," according to the latter document.[42] A more recent 2008 four-page brochure by the U.S. Geological Survey estimated the country's electric power generation potential from identified geothermal systems at 9,057 megawatts electric and from undiscovered geothermal resources at 30,033 megawatts electric. Additionally, a truly stupendous 517,800 megawatts electric could be generated by what the brochure described as "geothermal reservoirs in regions characterized by high temperature, but low permeability, rock formations," and, albeit less probably, perhaps another 200,000 megawatts electric.

At the end of 2009, the United States had a total installed capacity of 3,152.72 megawatts according to the Geothermal Energy Association (GEA), up from about 2,700 megawatts a decade earlier, making it the world's leader. Six new geothermal projects came on line that year, and 144 new plants were in development. The association said it was the beginning of accelerated growth that could add another 7,000 megawatts in coming years, aided by up to $338 million in Recovery Act funding that will support 123 projects in thirty-nine states, a "historic shift," according to a December 14, 2009, GEA release: "For the first time in over 25 years the DoE was putting significant resources behind its geothermal efforts."

Overall, geothermal energy ranks as the third largest renewable energy resource, behind hydroelectric power and biomass and exceeding solar and wind energy, according to estimates in the mid-1990s. Indeed, total geothermal resources are believed to be much larger than the combined resource bases of coal, oil, gas, and uranium. A November 2006 National Renewable Energy Laboratory technical report said, "Domestic resources are equivalent to a 30,000-year energy supply at our current rate for the United States."

Worldwide, 9,064.1 megawatts of geothermal generation capacity existed in 2005, the last year data were published by the Iceland-based

International Geothermal Association, up from about 7,000 to 8,000 megawatts a decade earlier. Geothermal power plants operate at high capacity factors (70–100 percent), and their availability factors typically exceed 95 percent. Globally, a belt of geothermal "provinces" rings the Pacific Ocean, curling southward from the Indian subcontinent through Southeast Asia to Indonesia, and then sweeping northward again along China's and Japan's shores, jumping to North America via the Aleutians and then stretching southward again along the western coasts of Canada, the United States, and South America. A smaller such "province" stretches from the Strait of Gibraltar across the Mediterranean Sea and into northern Italy, Yugoslavia, and Russia, with a southern extension hugging the Red Sea and reaching deep into eastern Africa. There are as well smaller geothermal hot spots in Iceland, the Canary Islands, and the Hawaiian Islands.

Geothermal energy is clean. According to that 1998 DoE brochure:

- No fuel is burned, there are no nitrogen oxide emissions, and sulfur dioxide emissions are very low.
- There are no air emissions with binary geothermal plants.[43]
- Many geothermal plants generate no appreciable waste.
- Plants have small land use over the project's lifetime and low impact in scenic regions.

They concluded that "geothermal energy is a vital part of a sustainable future."

6

Terra Transport: Hydrogen for Cars, Buses, Bikes, and Boats

"Basically, we can mass produce these now. We are waiting for the infrastructure to catch up."

That's what Kazuaki Umezu, the head of Honda's New Model Center, told reporters who in mid-2008 had come to cover a momentous event in the annals of hydrogen and fuel cell technology: the launch of the Japanese carmaker's—and the world's—first dedicated fuel cell car assembly plant in Takanezawa, a small town of some 30,000 residents about 80 miles north of Tokyo (figure 6.1).

Automotive assembly plants typically churn out hundreds of thousands or millions of cars, so the new plant is a modest affair and probably represents a learning tool as much as anything else.[1] Its launch task was to produce Honda's most polished, widely admired fuel cell passenger car, the Honda FCX Clarity. The plant's dedicated fuel cell vehicle assembly line includes unique operations not normally found in these types of plants, such as equipment for installing the fuel cell stack, manufactured at another nearby facility, Honda Engineering Co., in Haga-machi, and of the high-pressure compressed gaseous hydrogen fuel tank. The infrastructure issue—where and how do we get the massive hydrogen refueling capacity for the expected masses of hydrogen cars—remains a continuing concern in the United States, Europe, and Japan two years later but with some hope of resolution. This did not worry Umezu's boss, Honda's president Takeo Fukui (he was succeeded in early 2009 by Takanobu Ito). He told another reporter that about a hundred years ago, Ford's Model T came out, kicking off the development of the auto industry: "If you ask 'were there any gas stations back then?' there weren't. In the car industry, cars should come first, and the infrastructure will follow." He then conceded that it may take some time, and perhaps ten

Figure 6.1
Honda technicians install the hydrogen fuel tank during assembly of the company's fuel cell Clarity passenger car at the company's ground-breaking New Model Center in Takanezawa, Japan, in summer 2008. The center houses the world's first dedicated fuel cell assembly line production facility for limited-series production of fuel cell vehicles.

years, to get to "some level of infrastructure." Introduced as a concept mock-up at the 2005 Tokyo Motor Show, it was a finished product ready for automotive writers for test driving by the time it reached the November 2007 Los Angeles Auto Show. Honda planned to build about 200 Claritys in three years for delivery to customers in Japan and the United States. The Honda assembly plant debut and the Clarity launch as a lease car to selected customers, including high-visibility Hollywood celebrities such as actress, author, and environmental activist Jamie Lee Curtis, were important milestones in the advancement of hydrogen and fuel cell technologies during the past two decades. These environmentally beneficial, clean transportation technologies were moving, if gradually and slowly, to mainstream industrial status.

It had been a long transition from Daimler-Benz's first fuel cell vehicle, the cobbled-together, rudimentary 1994 NECAR 50-kilowatt proton exchange membrane (PEM) fuel cell van with a range of 130 kilometers

(81 miles): just about the entire cargo space was filled with fuel components, testing equipment, and other gear, barely leaving space for the
driver and a passenger. By 2010, entire fleets totaling hundreds of cars
were already on the road in the United States, Europe, and Japan, with
many of them loaned or leased to ordinary commuters. General Motors's
Project Driveway fleet of more than 100 fuel cell Equinox SUVs (115
kilowatt-hours with a range of 200 miles), for instance, has already
racked up millions of miles since its public launch in November 2007.
As noted, Honda is deploying some 200 FCX Claritys (100 kilowatt-
hours with a range of about 270 miles), Mercedes-Benz started building
a fleet of 200 Mercedes-Benz B-Class F-Cell sedans (100 kilowatt-hours
with a range of about 250 miles; figure 6.2), also destined largely for

Figure 6.2
Mercedes-Benz's latest fuel cell passenger car, the B-Class F-Cell. Small-volume
production of about 200 cars began in fall 2009, and they became available for
lease to a few customers in California in late 2010 at $849 per month, including
the cost of hydrogen fuel and insurance. The picture shows the car at a hydrogen
fueling station at the airport in Stuttgart, Germany, the carmaker's home town.
The sign in the background, above the car's hood, shows the hydrogen (Wasserstoff) price: 0.9 euros per 100 grams, or 9 euros ($12.07) per kilogram; 1
kilogram of hydrogen has roughly the same energy content as 1 gallon of
gasoline.

average drivers, near the end of 2009, and Toyota was fielding about 100 Toyota FCV SUVs (90 kilowatt-hours with a range of about 431 miles in a 2009 road test) in California, Japan, and Germany. Hyundai/Kia was launching eighty of its latest-generation Tucson iX SUVs in South Korea, and Germany's Volkswagen has shipped twenty-two copies of its China-made Lingyu cars to California, with a few more to be deployed in Germany.

Around the World in 125 Days The Mercedes Fuel Cell World Tour: in 2011, Mercedes-Benz sent a strong message that fuel cell cars were ready for prime time by sending a three-car convoy on a drive around the world. The marching orders for the three B-Class F-Cell vehicles were to circle the globe in 125 days, starting in Stuttgart at the company's headquarters at the end of January 2011. The symbolic idea was to commemorate the 125th anniversary of the birth of the automobile in 1886, the year when the man who gave the company its name, Karl Benz, registered his patent for his "vehicle powered by a gas engine." Sent off by German chancellor Angela Merkel, the three light-green Mercedes F-Cell electrics started out going west on the F-Cell World Drive, a four-continent, fourteen-country 30,000-kilometer (18,750-mile) journey that crossed France, Spain, and Portugal on the first stage. From Lisbon, the cars were airlifted to Miami for the drive across the southern United States and north to Vancouver in Canada, from where they were flown to Sydney. After driving across Australia to Perth, they were air-shipped to Shanghai and driven the rest of the way across China, Kazakhstan, Russia, and northern Europe, back home to Stuttgart. En route, each fuel cell car, which can cover 400 kilometers (250 mile) on one tank fill, was refueled about 130 times by hydrogen dispensed from a support vehicle, a Mercedes-Benz Sprinter van modified by industrial gas supplier Linde AG with a 700 bar (10,000 psi) compressor/dispenser. The cars and their support convoy—the refueler, staff SUVs, a boom-equipped video van—arrived back at company headquarters in Stuttgart on June 1, with no reported problems with the fuel cell technology. The only significant visible damage was a dented left rear door and fender following a collision in Kazakhstan when a Russian Lada car rocketed out of a side street and bashed one of the three Mercedes cars, requiring the replacement of some rear suspension components. The press conference following the arrival provided a high-visibility platform for Dieter Zetsche, CEO of the parent Daimler company, to announce the company's next step, construction of 20 new hydrogen fueling stations in Germany together with Linde at a total cost of 20–25 million euros (see infrastructure issue, discussed later in this chapter). And in a second significant announcement, Daimler's top r&d executive, Thomas Weber, announced that the company was moving up the start of series production of fuel cell cars one year—from 2015 to 2014.[2]

In the city transit bus category, Mercedes-Benz, starting in 2003, led the way with the deployment of, ultimately, thirty-six fuel cell versions of its bread-and-butter internal combustion–engined Citaro buses across Europe and in China and Australia. Most of them have been retired by now, although six were still operating in Hamburg in 2010. Ten new ones, laid out as hybrids (the original ones used straight fuel cells) were destined for Hamburg in 2010, with another ten scheduled for delivery in 2013, and more expected after 2018 with a unique procurement program. The first of four large, articulated Dutch-German Phileas fuel cell hybrid buses, at 18 meters (59 feet) and 105 passengers, described as the world's biggest, made its debut at the 2010 World Hydrogen Energy conference in Essen, Germany, with all of them scheduled to see regular service in Cologne and Amsterdam. London's transit authority was scheduled to receive eight fuel cell hybrid buses by 2012 as part of a European Union Lighthouse project, with many more scheduled to be delivered to Hamburg, Turin, Milan, Oslo, and probably other European cities. A triple hybrid bus was announced in summer 2009 by Czech trolley maker Skoda Electric (not to be confused with carmaker Skoda, which is part of the Volkswagen group) and German fuel cell developer Proton Motor Fuel Cell. The battery/ultracapacitor combination is said to be more efficient and offers more effective power recovery with regenerative braking, the designers of a similar "tribrid" bus at Glamorgan University in Wales said when they announced their project at a December 2007 electric vehicle symposium in Anaheim, California. The bus itself was shown at the Grove Fuel Cell Symposium in London in September. Ten, or maybe more, Belgian-built Van Hool hybrid fuel cell buses powered by American UTC fuel cell power plants were on order for AC Transit in the San Francisco Bay Area, and Latin America's first fuel cell bus started tests in late summer 2009 in São Paulo, Brazil, with three more, all built in Brazil by manufacturers Marcopolo and Tut-totrasporti with fuel cell engines coming from Canada's Ballard Power Systems and Germany's Nucellsys (a once- or twice-removed Ballard-Daimler offspring), to be added to the test fleet by 2011.

An unusual vehicle freighted with historical symbolism was unveiled in February 2009 in Turin, Italy. Fiat's agricultural machinery New Holland subsidiary rolled out a funky 75 kilowatt-hour NH^2 hydrogen fuel cell tractor based on its existing T6000 tractor series (figure 6.3). Looking a little like an oversized Tonka toy, New Holland said it may be the wave of the future for farmers who would generate their own fuel, or also ammonia widely used as fertilizer. It was probably a coincidence, but the rollout came exactly fifty years after the world's first fuel cell tractor, built by what was then Allis-Chalmers, was photographed plowing a midwestern field—by some accounts, the first fuel cell–powered vehicle ever.

Figure 6.3
Photographers and reporters swarm around a fuel cell–powered tractor during its official rollout in Turin, Italy, in February 2009. The vehicle was converted from a conventional model by New Holland, the Fiat-owned farm equipment subsidiary.

Fuel Cells at Sea, and Below

The biggest fuel cell–powered vehicle is mostly out of sight, literally cruising below visibility: it is a class of hydrogen-fueled, fuel cell–powered stealth submarines developed by the German navy and the Howaldtswerke-Deutsche Werft AG (HDW) shipyard in Kiel (figure 6.4).[3] Begun in the mid-1990s, the 212/214-Class submarine uses a conventional 3.96-megawatt twin-diesel power plant for normal cruising at speeds of up to 20 knots (23 miles per hour) submerged. A paper presented in 2006 at the Advanced Naval Propulsion Symposium in Arlington, Virginia, by the Siemens scientist principally responsible for the fuel cell's design, Albert E. Hammerschmidt, said the 57-meter 212 version uses nine 34-kilowatt fuel cells for a total of 306 kilowatts. The 65-meter 214 version intended for export uses two more efficient 120-kilowatt modules for a total of 240 kilowatts. Hydrogen is carried in metal hydride canisters, and oxygen is carried as a liquid. Some of the

Figure 6.4
German sailors stand at attention during the 2002 christening ceremony of one of the first of a series of fuel cell–powered stealth submarines designed and built by the German Howaldtswerke-Deutsche Werft shipyard. The 300-kilowatt PEM fuel cell was designed and built by Siemens.

operational specs are still secret, but a Wikipedia entry says that when in stealth mode, the PEM fuel cells drive the ship at speeds estimated at 2 to 6 knots (2.3–6.9 miles per hour), with a maximum submerged fuel cell range of 780 kilometers and an energy-conserving speed of 7 kilometers per hour (4.3 miles per hour). On the surface, the fuel cell power range is 2,310 kilometers (1,444 miles); with the main diesel engines running, the range is 12,000 miles surfaced and 487 miles submerged. Variants of the ship have been delivered to the Greek, South Korean, Pakistani, and Turkish navies.

Fuel cells are also powering a couple of inland waterway tourist cruise boats in Holland and Germany. The most recent example is the Dutch *Nemo H2*, which was christened and launched in December 2009 in Amsterdam by Boat Company Lovers (the boat operators). The 22-meter vessel can carry eighty-six passengers and is propelled through Amsterdam's canals by stern and bow thrusters using two 30-kilowatt PEM fuel

Figure 6.5
What was described as the world's first fuel cell–powered inland waterways tourism ship, the 72-ton *Alsterwasser* Zemship (zero emissions ship), is docked in downtown Hamburg awaiting its first passengers in fall 2008.

cells linked to a 70-kilowatt-per-hour battery. A somewhat larger Zemship (zero emissions ship), the 100-passenger *Alsterwasser*, got underway in Hamburg in 2008 (figure 6.5). Propulsion is provided by two 48-kilowatt PEM fuel cells built by Proton Motor Fuel Cell. In addition, a number of small fuel cell systems have been installed and tried in recent years in motorboats and for auxiliary sailboat power in California, Austria, Germany, and presumably elsewhere. Other recent examples are fuel cells as power plants of choice for ships and boats. Six fuel cells built by Canada's Hydrogenics were ordered in early 2010 for a boat project in Istanbul, Turkey, coordinated by UNIDO's International Centre for Hydrogen Energy Technologies (ICHET) there, a follow-up to four fuel cell systems for boats ordered the previous year for Istanbul's Municipal Transit Authority.

On display during the 2009 Copenhagen Climate Summit, a big fuel cell was part of a ship docked at a city pier for a few days: the 5,900-ton *Viking Lady* (figure 6.6). This bright orange-and-yellow offshore supply ship operated by Norwegian fleet operator Eidesvik for the Total oil

Figure 6.6
The 5,900-ton *Viking Lady* offshore supply ship equipped with an auxiliary molten carbonate fuel cell. It was docked in Copenhagen in December 2009 for the Copenhagen Climate Summit.

company, is equipped with a 320-kilowatt-hour molten carbonate fuel cell built by MTU to provide onboard auxiliary power from liquefied natural gas, not hydrogen.

All these are follow-ups to projects dating back half a century that began with conventional internal combustion–engined cars converted to hydrogen fuel in the 1960s and 1970s. The first hand-built fuel cell cars appeared in the late 1980s. Beginning in 2006, BMW built a fleet of big, elegant, switchable-on-the-fly, dual-fuel gasoline /liquid hydrogen–fueled internal combustion sedans dubbed BMW Hydrogen 7 (191 kilowatts and 260 horsepower, with more than 125 mile range on hydrogen and 312 miles on gasoline), mostly for VIPs and celebrities, which in 2010 were expected to be gradually phased out. Fourteen of them did VIP-ferrying duty at the misbegotten Climate Summit in Copenhagen, and other manufacturers, like Mercedes, Honda, GM, Fiat, Volvo, and the tiny Scandinavian Think Hydrogen car, had their cars there as well. Not far behind, and seemingly more aggressive, is Korea's Hyundai: media and competitors say Hyundai plans to produce fuel cell cars by the thousands by 2012, primarily for the Korean domestic market at the start.

If these optimistic projections in fact come true—some people remember the 1990s hype by respected industry spokespeople proclaiming that thousands of fuel cell cars would be on the road by the mid-2010s—Hyundai will not be alone for long. Other carmakers, including Mercedes, Honda, Toyota, and General Motors, are targeting 2015 as the breakout year when they will begin marketing fuel cell cars in probably limited numbers—and probably high prices initially—to standard consumers.

At the start of the new century's second decade, an estimated 740 cars and more than 50 buses are or will be operating on roads almost exclusively in developed countries, made by manufacturers both to gain experience and data from ordinary drivers and for public relations purposes—a huge jump from the 50 or so vehicles that were believed to be operating in 1995. In addition, some 400 prototypes and experimental cars are listed on the interesting and instructive Web site by the German TÜV Industrie Service, "H2 Mobility: Hydrogen Vehicles Worldwide" (www .netinform.net/H2/H2Mobility/Default.aspx) whose time line goes back to the early eighteenth century. Among the gems are a 1941 Russian

GAZ-AA pickup truck, a 1933 Norsk Hydro pickup truck with an onboard ammonia reformer, an 1860 Lenoir Hippomobile, and an 1807 experimental vehicle built by French inventor François Isaac that covered 100 meters, propelled by an internal combustion engine with hand-triggered ignition and fueled by hydrogen gas carried in a balloon.

Fuel Cell Two- and Three-Wheelers

A fuel cell–propelled utility tricycle dubbed Cargobike was introduced in 2005 by German manufacturer Masterflex. A 250-watt PEM fuel cell mated to an electronic transmission, plus the occasional pedal power, provides a leisurely top speed of about 18 kilometers per hour (11 miles

Figure 6.7
A service technician for phone company Deutsche Telekom en route to a customer on his fuel cell–powered Cargobike in Berlin. Deutsche Telekom began testing fourteen of them in early 2008. With a maximum speed of 25 kilometers per hour (16 miles per hour), they are classified as bicycles; they do not require a driver's license and are allowed to operate on bicycle paths.

per hour) with a maximum load of 150 kilograms. It is one of a half-dozen small fuel cell vehicles and devices—scooters, wheelchairs, utility vehicles, midibuses, portable generators—that are being tested in small fleets in four European regions in the HYCHAIN project with support from the European Union and with the goal of eventual commercialization. These Cargobikes, classified as a bicycle, do not require a driver's license; service technicians and mail carriers, for example, can use bicycle paths for their rounds.

Also, more than three dozen fuel cell scooter or motorcycle projects have been started, and frequently abandoned, over the years. One of the latest, with perhaps a real chance of making it to market, is the Burgman. Despite its Teutonic-sounding name, it is a joint project by Japan's motorcycle maker Suzuki and the U.K.'s fuel cell developer Intelligent Energy. The 125-cc-class scooter was originally shown at the 2009 Tokyo Motor Show and publicly launched in London in February 2010. Wrote British

Figure 6.8
A hybrid fuel cell–powered scooter developed jointly by the U.K. fuel cell developer Intelligent Energy and Japan motorcycle manufacturer Suzuki. Launched at the 2009 Tokyo Motor Show, this picture shows the scooter during its European debut in London in early 2010.

motor journalist Guy Procter in *Motorcycle News,* "History has rarely been made with less fanfare. . . . It isn't the stuff of petrolhead dreams, but it's right on the money for a 125-class commute. . . . Its styling won't win you Prius-like glances of approval from wan eco-chicks. But that, says Suzuki, is partly the point."[4] Its makers did not say anything publicly about commercial availability, but some media reported it might hit salesrooms in 2015, increasingly regarded or, better perhaps, hoped-for, as Year One of the hydrogen transport age.

Another Intelligent Energy–powered vehicle is the fuel cell version of the venerable London taxi. It is supposed to appear on London streets

Figure 6.9
A fuel cell version of the traditional London taxi. Henri Winand, CEO of Intelligent Energy, left, stands with Kit Malthouse, London's deputy mayor for policing and chair of the London Hydrogen Partnership, in front of the prototype of the cab unveiled in June 2010. A small fleet of the new clean cabs is ined to operate in London in time for the 2012 London Olympics.

in time for the 2012 Olympics. The black cab, hybridized with lithium-ion batteries, has a range of more than 250 miles on a tank of hydrogen, can get up to more than a respectable 80 miles per hour, and is refueled in about five minutes. In addition to Intelligent Energy, its team included Lotus Engineering consultancy, taxi manufacturer LTI Vehicles, and product developer TRW Conekt. Plans for the cab's fuel cell version were first announced in early 2008. Intelligent Energy said in 2010 the first, presumably small, fleet of fuel cell cabs will be launched in London in 2012, in time for the London Olympics.

Hydrogen Fueling Infrastructure: The Chicken-and-Egg Issue

All of these hoped-for vehicles are predicated on the availability of hydrogen fuel in a widespread hydrogen fueling infrastructure—a big, and so far, unresolved problem. This chicken-and-egg problem—What comes first: fleets of hydrogen cars or a fueling infrastructure?—is beginning to be tackled in earnest in Germany and Japan, but also elsewhere. According to a worldwide list of hydrogen stations on the Web site of the American nonprofit Fuel Cells 2000 (www.fuelcells.org) dated November 2009, more than 200 fueling stations of all types—stationary or mobile; for compressed hydrogen gas at 5,000 or 10,000 psi; liquid hydrogen—existed worldwide, most of them with restricted access (only for the companies or institutions that operated them) and few of them accessible to the general public. Another 67 were in various planning phases. These numbers largely agree with the tally maintained by Ludwig-Boelkow Systemtechnik, a German consultancy active for decades in hydrogen and fuel cell energy, which says 206 stations are operating worldwide and 116 are being planned (slightly more than 50 have already been shut down). The United States accounted for about 86 operating stations, including 38 in California. Canada counted 10. Europe's cheerleader, Germany, has built 29 of them, but only 4 were open to the general public, according to a presentation by Daimler's manager for fuel cell vehicles, Ronald Grasman, during a February 2010 Webinar organized by the National Hydrogen Association. Japan had 24 stations, and Korea had built 6. Small Denmark had put up 6 stations, and giant China had 3. There were 2 in Singapore, 1 each in and Brazil, and none in Africa.

Past efforts to set up widespread fueling facilities have been hampered in part by what some fuel cell partisans regarded as foot dragging on the part of the energy companies and government. That view was exemplified by critical comments from former General Motors R&D vice president Larry Burns at the National Hydrogen Association's 2008 annual conference. "We have now reached a point where the energy industry and governments must pick up their pace so we can advance in a timely manner," said Burns in his prepared remarks. "GM remains committed to continuing this journey as long as we see clear evidence that the energy industry and governments are committed to developing a hydrogen infrastructure. And we need to see this evidence soon. . . . Quite frankly, it makes little sense for us to allocate resources to fully develop and deploy a vehicle beyond our current Equinox Fuel Cell if we are not confident energy suppliers will provide sufficient refueling stations."[5]

A year later, the issue was being tackled with evident seriousness of purpose in Germany and, perhaps even more so, in Japan. (The United States was still dragging behind, partly because of the high visibility of and relentless PR for battery electrics and plug-in hybrids, reinforced by the perception that the Nobel Prize–winning secretary of energy, Steven Chu, was not a champion of hydrogen and fuel cells for transportation.) That fall, six major industrial companies—carmaker Daimler; energy companies Vattenfall and EnBW; oil and gas companies Total, Shell, and Austria's OMV; plus the German national hydrogen and fuel cell organization NOW as coordinator—signed an "H2 Mobility" memorandum of understanding to evaluate and push for a national hydrogen infrastructure. And in what looked like a well-orchestrated one-two punch, the September 10 announcement followed by two days the release of a letter of understanding signed by top executives of carmakers Daimler, Ford, General Motors, Honda, Hyundai/Kia, Renault/Nissan, and Toyota pledging to implement production and commercialization strategies for launching large numbers of fuel cell cars by 2015. The letter referred to about "a few hundred thousand over life cycle on a worldwide basis," but that may be a bit too optimistic. One American observer closely attuned to the international fuel cell scene liked what he saw but felt that "the numbers may be vastly inflated." Still, he thought this was a serious effort on the part of Germany AG: "This time the government appears to be stepping in to help with infrastructure. Also, the technology is far superior by now, and costs are plummeting." The H2 Mobility paper was endorsed by Germany's transport minister Wolfgang Tiefensee, who supports automotive electrification in general as a climate-friendly necessity and hydrogen energy in particular (he was being chauffeured around Berlin in a hydrogen car, German media reported). "Today, we can see that Germany is setting the pace when it comes to hydrogen and fuel cell technology," he said. "We are aiming at establishing a nation-wide supply with hydrogen in Germany around 2015 in order to support the serial production of fuel cell vehicles."[6]

The German announcements came a couple of months after reports from Tokyo that Japanese energy companies were banding together to push the creation of a hydrogen fueling infrastructure. An August 4, 2009, *Nikkei* story said that thirty energy companies were joining in a group called the Research Association of Hydrogen Supply/Utilization Technology to conduct "joint research with an aim to commercialize technologies for supplying hydrogen to fuel cell vehicles by fiscal 2015." The group includes oil companies Nippon Oil, Idemitsu Kosan, Cosmo Oil, Japan Energy, and Showa Shell Sekiyu; gas utilities Tokyo Gas, Osaka Gas, Toho Gas, and Saibu Gas; and industrial gas suppliers and builders of hydrogen stations Iwatani, Taiyo Nippon Sanso, Air Liquide Japan, and Mitsubishi Kakoki Kaisha, reported Hirohisa Aki, deputy director for technology development in the new and renewable energy policy division of the Ministry of Economy, Trade and Industry (METI) on December 2, 2009, at the Twelfth IPHE Steering Committee meeting in Washington, D.C. The research alliance will conduct field trials by setting up "dozens of hydrogen stations" across Japan, according to the *Nikkei* story, and by using the oil companies' hydrogen production facilities and the pipelines of the gas companies, the group will research ways to "transport the fuel to filling stations in a stable manner at low cost." The *Nikkei* story said the group will function as the infrastructure voice to deal with carmakers and the government. It is funded by METI, and the member companies are providing twenty executives and experts as full-time staff to the new alliance.

Next, infrastructure moved to the Europe-wide stage. In November 2010, a major study, organized by McKinsey & Company and based on data from thirty-one companies and organizations—carmakers, oil and gas companies, utilities, industrial gas companies, equipment manufacturers, nongovernmental and governmental entities—was presented in Brussels calling for a study of a Europe-wide market launch plan for fuel cell electric vehicles and hydrogen infrastructure and a coordinated rollout of battery-electric vehicles, plug-in hybrid electric vehicles, and a battery-charging infrastructure, probably taking its cue from the German model. McKinsey had been commissioned to provide support for analyzing the data provided by these groups, and the findings were expected to be reflected in upcoming European Union Commission policy papers.

All of these initiatives—Honda's prototype manufacturing plant, improved and more durable vehicles, fleet tests, infrastructure development in Europe and Japan—are the latest examples of efforts that were underway for approximately the last three decades of the twentieth century and, as noted above, the first decades of the twenty-first. Europe's first quasi-commercial hydrogen fueling station that in effect served only one vehicle, a delivery van, opened in the German port city of Hamburg in January 1999, the tail end of what was probably hydrogen's biggest decade so far in terms of hope and public expectations. The van, a 3-ton, 2.3-liter, 4-cylinder internal combustion engine Mercedes-Benz, belonged to a delivery fleet operated by the mail-order company Otto-Versand, one of the world's largest. Otto-Versand had been promoting environmental responsibility and sustainability in its products for years. It began experimenting with natural gas-fueled vans in 1995, and in 1998 it installed a 50-kilowatt photovoltaic system, then the biggest in northern Germany, at one of its facilities. In 1997 it teamed up with a dozen local companies on the hydrogen gas station project initiated by the Hamburg Hydrogen Society, a nonprofit group founded in the 1980s.

The participants in the project included two banks, Hamburg's transit authority, the city's electric and gas utilities, a gas distributor, a specialized moving company, and, notably, Deutsche Shell AG, the German division of the Royal Dutch/Shell Group. The van's conversion to hydrogen operation was designed, built, and tested by a veteran American hydrogen specialist, Frank Lynch of Hydrogen Components in Littleton, Colorado.

In addition to wide coverage in regional and national papers, all of Germany's TV stations and CNN showed footage of the van being refueled and driven in and out of the Ludwig Boelkow Fueling Station, named after the late aerospace industrialist who was the patriarch among Germany's hydrogen supporters (Boelkow died in 2003). Even the *New York Times* carried a brief story on January 13, 1999: "Hydrogen will be the most important energy source of the 21st century," Fritz Vahrenholt, a top Deutsche Shell executive at the time, was quoted as saying. "Long term, it will replace oil and gas."

Two months later, in Washington, D.C., DaimlerChrysler made headlines around the globe with the unveiling of its latest fuel cell car, NECAR (New Electric Car) 4. DaimlerChrysler cochairman Bob Eaton drove the tiny four-seater onto center stage at a jammed press conference in the Ronald Reagan International Trade Center. The liquid hydrogen–powered NECAR 4 (which was based on the A-Class four-passenger subcompact, sold only in Europe) had a top speed of 90 miles per hour and a range of almost 280 miles. The twin fuel cell stacks put out 70 kilowatts. (The earlier versions, the 1996 NECAR 2 and the 1997 NECAR 3, are described later in this chapter.)

Lest anyone doubted that DaimlerChrysler was into fuel cells for the long term, cochairman Jürgen Schrempp said at the Washington festivities: "Today, we declare the race to demonstrate the technical viability of fuel cell vehicles is over. Now, we begin the race to make them affordable." The company planned to have fuel cell vehicles in limited production by 2004, Schrempp said—a boast that did not materialize. Toward that end, it intended to spend more than $1.4 billion on fuel cell technology, he added—about the same amount of money invested in an entire line of profit-making vehicles.

Michael Otto, the CEO of the mail order house bearing that name, and Schrempp were emblematic of the change of attitude among some international business leaders who came out in favor of advanced hydrogen-based energy technologies as the twentieth century drew to a close. Other key executives said about the same. "Looking at the total system, including its packaging and incorporating the fuel cell and hydrogen absorbing alloy, we've achieved what I believe to be the world's highest standard," said Yoshio Kimura, Toyota's general manager for the experimental fuel cell RAV4 project at the car's unveiling at the Osaka International Electric Vehicle Symposium in October 1996. And John Smith Jr., chairman of the board of directors of General Motors, predicted at the 1997 Detroit Auto Show that "environmental pressures will force changes in the automotive industry and lead to a fundamental shift away from gasoline-powered vehicles to electric, hybrid, and fuel-cell-powered vehicles."

This was a profound change in attitude. Through the 1970s and probably until the late 1980s, any talk of advanced propulsion technologies, clean energy, or concern with environmental issues was considered heresy in the blinkered automotive world of Detroit. William Clay Ford Jr. got a taste of that when he tried to raise environmental issues after joining the company founded by his great-grandfather, Henry Ford, as a vehicle planning analyst in 1979. "Coming to an old-line auto company, people looked at me like I was a Bolshevik for bringing it up, for even asking the questions," Ford told a *New York Times* reporter in early 1999.[7]

The case for clean alternative transportation fuels in general and hydrogen in particular has been argued for years in all sorts of forums. The intent here is not to plow the same ground again in great detail. Still, some numbers are pertinent.

In 2008, the last year for which data were available from the U.S. Energy Information Administration (EIA—December 2009 report) at the time of writing, transportation accounted for 27.8 percent of total primary energy consumption in the United States (electric power generation was the biggest user, with 40.1 percent). Petroleum was the biggest single primary energy source for the country, with 37.1 quadrillion Btu (quads) or 14.72 million barrels of crude oil per day (4.96 million domestic production, 9.76 million barrels imported). Of that total, 71 percent went to transportation (23 percent to industrial use, 5 percent to residential or commercial, and 1 percent to electricity production). Total U.S. greenhouse gas emissions (in addition to carbon dioxide, methane, nitrous oxide, and a catch-all category of gases with high global warming potential—hydrofluorocarbons, perfluorocarbons, and sulfur hexafluoride) in 2008 were 2.2 percent lower than in 2007, mostly because of higher energy prices, especially during the summer driving season, which led to a drop in petroleum consumption; economic contraction in three out of four quarters, resulting in lower energy demand; and lower demand for electricity combined with lower carbon intensity of electricity supply.

U.S. carbon dioxide (CO_2) emissions in 2008 were estimated to total 5,839.3 million metric tons, down from 6,017 million metric tons the previous year, but significantly higher than the earliest figure in that EIA chart, 5,022.3 million metric tons in 1990. Petroleum was the largest single fossil fuel source for energy-related CO_2 emissions, accounting for 41.9 percent (coal is second largest at 36.5 percent, although it produces more CO_2 per unit of energy than other fossil fuels), but petroleum consumption accounted for 44.6 percent of total U.S. consumption in 2008 compared to coal's 26.8 percent. Natural gas, with a carbon intensity 55 percent of that of coal and 75 percent of that of petroleum, accounted for 28.5 percent of U.S. fossil energy use but only 21.4 percent of total energy-related CO_2 emissions. A chart for worldwide emissions through 2006 only—later data were not yet available in this interim report—show that in North America, the United States was the biggest producer of CO_2: 5,903 million metric tons out of 6,954 million metric tons for the region. Globally, the United States was overshadowed by China, with 6,018 million metric tons (out of a total of 11,220 million tons for Asia and Oceania). In marked contrast, economic powerhouse Japan belched out a relatively modest 1,247 million metric tons that year. And all of Europe emitted less than the United States: 4,721 million metric tons. The world's grand total for that year was estimated at 29,195 million metric tons, significantly more than the CO_2 output in 1997, the earliest year on that chart, with 23,247 million metric tons.

Transportation is the second largest source of CO_2 emissions after electric power production: it accounted for 33.1 percent of the total (motor gasoline, diesel fuel, and jet fuel), and electricity production accounted for 40.6 percent of the U.S. total that year, according to the EIA data.

A Key Issue: How to Store Hydrogen on a Vehicle

Hydrogen is widely regarded as the ideal or nearly ideal fuel to solve these problems. By definition, it does not pollute. Burned or oxidized with atmospheric oxygen, it produces water. (It also produces some nitrogen oxides if combustion occurs with a flame, as in an internal combustion engine, but no nitrogen oxides are produced in an electro-chemically reactive fuel cell.)

The difficulty lies in how to carry hydrogen onboard a vehicle. In gasoline- or diesel-powered cars, storing conventional liquid fuel is easy. Conventional front-engine designs usually have the gas tank in the rear, beneath the trunk. With hydrogen, the situation is more complicated. In its ambient state, hydrogen is a gas. For efficient storage, it must be compressed like natural gas, cooled into a cryogenic state, bound within the structure of a hydride, or put into some other form (e.g., a slurry). All of these are difficult or have engineering or economic drawbacks (or both), especially on a space-constrained passenger car. Carmakers and researchers have wrestled with this for decades and are still at it. One early Chinese station wagon solved the problem by placing compressed hydrogen tanks on its roof (a picture of the car was shown at the 2002 World Hydrogen Energy Conference in Montreal by Japanese researcher Kazukiyo Okano who had taken it earlier in Beijing; figure 6.10). Roof storage is still used on transit buses, but now with the tanks artfully camouflaged by sheet metal. Both Toyota and Mercedes have tried storing it in metal hydrides, but they are heavy, and the idea has been pretty much discarded. Another tack has been to extract it from a carbonaceous fuel like gasoline or methanol, but both have been abandoned as too complicated—you need a "chemical plant under the hood" was a dismissive description. BMW has steadfastly supported supercold cryogenic liquid hydrogen storage, including a recent variant known as a cryogenic-capable system it has developed in collaboration with researchers at Lawrence Livermore National Laboratory.

Today the preferred method of onboard hydrogen storage is compressed gaseous at pressures of, so far, 5,000 and 10,000 psi (350 and 700 bar); even higher pressures have been tried in experimental tank designs, but as far as is known, no carmaker has put one into a real-world hydrogen car to be driven on public roads. Compressed-gas tank designs employing increasingly lighter and tougher carbon fiber and composite materials have made it possible for carmakers to achieve and exceed the benchmark range of 300 miles, deemed essential for marketing advanced-technology cars to consumers. Each of a fleet of ten Toyota Fuel Cell Hybrid Vehicles—Advanced (FCHV-adv) deployed in 2009 and 2010 at New York's JFK airport as airport staff cars achieves a range of about 350 miles before refueling. One Toyota of this type even covered an estimated 431 miles on a single tank of 700 bar compressed

Figure 6.10
An early Chinese ZEV fuel cell–powered station wagon, described by a Japanese researcher at the 2002 World Hydrogen Energy Conference in Montreal. The conventional-looking car was converted to hydrogen by Beijing Green Power Co. Compressed hydrogen was carried in three rooftop-mounted compressed hydrogen tanks.

hydrogen in a 2008 test conducted with the U.S. Energy Department and two national laboratories, averaging the equivalent of 68 miles per gallon.

In the 1970s, when people first thought about using hydrogen as fuel for cars and trucks, many assumed that heavy steel-walled pressure bottles, similar to those used in welding, would be used to store the fuel onboard. Such bottles are simple, but their weight would be prohibitive. The late Larry Williams of Martin Marietta Aerospace once calculated that a conventionally constructed steel pressure tank capable of holding roughly the same amount of energy as the fuel tank of a standard-size car would have to weigh about 3,400 pounds, would require a pressure of 800 atmospheres, and would have to have steel walls almost 3 inches thick.

Williams was a strong early advocate of cryogenic liquid hydrogen storage, foreshadowing BMW's decades-long practice. He listed the following main advantages in a 1973 paper:

- Lowest cost per unit energy
- Lowest weight per unit of energy
- Simple supply logistics
- Normal refuel time required
- No unsurmountable safety problems

He then listed these disadvantages:

- Loss of fuel when vehicle is not in operation (the supercold liquid evaporates)
- Large tank size
- Cryogenic liquid engineering problems

Some of the problems that Williams listed—large tanks, engineering problems, boil-off—have been progressively minimized by BMW in two decades of development work. As early as 1996, a BMW paper illustrated progress with photographs of the tanks in the trunks of BMW's four generations of liquid hydrogen–fueled cars. The first picture of the original 1979 model showed a tank, with protruding pipes and fittings, occupying most of the trunk space. The 1984 version was a smoother ovoid-shaped structure that still occupied much of the space. The 1990 version was a smaller cylindrical structure that left much more room for luggage. In the 1995 version, the tank was so intelligently packaged and incorporated in the trunk above the rear axle as to be almost invisible, leaving lots of useful space.

A further refinement is the next-generation so-called cryo-compressed hydrogen onboard storage that BMW has been working on with a Lawrence Livermore team headed by Salvador Aceves. As the name implies, it combines characteristics of highly insulated supercold liquid hydrogen storage with advanced high-pressure tank technology. In a paper presented at the 2010 National Hydrogen Association annual conference in Long Beach, BMW researcher Klaas Kunze said this new technology offers a lot of refueling flexibility with different hydrogen states, from low-density compressed gaseous hydrogen (24 grams per liter at 5,000 psi) or high-density cryo-compressed hydrogen (65–70 grams per liter at

3,500–4,200 psi). In effect, it permits refueling from liquid hydrogen stations offering high vehicle range and fast, cost-efficient refueling while pressurization of the gradually evaporating liquid keeps the gaseous boil-off hydrogen safely inside the tank for much longer periods of time—possibly weeks. "Cryo-compressed storage has been identified as most promising complement to the existing hydrogen storage portfolio, in particular for use in larger passenger vehicles with high energy requirement," said Kunze in his abstract. The goal is to build a fundamental proof-of-concept piece of equipment by 2011.

For a long time, hydrides were considered to be perhaps the best solution to the problem of how to store hydrogen in an automobile. Safety was the main reason. Ever mindful of the Hindenburg syndrome, researchers generally consider hydrides safer all around, because they cannot spill or vent hydrogen and do not burn in a crash. In recent years, however, the weight penalties associated with hydrides have dampened the enthusiasm for them considerably

Hydrides, usually metal alloys, were originally developed for nuclear power plants to slow down fast neutrons as a means of controlling nuclear reactions and thus controlling the output of a nuclear power plant.[8] Titanium-iron hydride, the material used in early automotive applications, looks and feels like any ordinary metal—tiny silvery granules without any hint of anything unusual. However, the granules have the remarkable characteristic of absorbing hydrogen in an almost sponge-like fashion, with different hydrides soaking up and releasing the gas at different pressures and temperatures.

Heat is given off when the alloy or other storage material absorbs hydrogen (heat of formation); the same amount of heat must be added to the hydride to release the hydrogen again (heat of decomposition). Hydrides bind the hydrogen atom atomically; the hydrogen atom is integrated into the crystalline structure of the hydride, with the hydrogen atom's electron transferred to the hydride.

Integrated into the alloy storage material, hydrogen does not take up any additional volume, as a gas or a liquid would. For this reason, any hydride can carry much more hydrogen energy in terms of volume than liquid hydrogen, but there is a huge weight penalty. For example, a 100-liter tank of titanium-iron hydride carries 1.2 to 1.5 times as much energy as a 100-liter tank of liquid hydrogen, but it weighs about twenty-five times as much.

Methanol

The benchmark for efficient onboard energy storage is still gasoline. Because it's a liquid, something consumers are familiar with, methanol was championed for a while by some experts as a logical hydrogen "carrier" and cleaner successor to gasoline. The average tank, full of gas, weighs about 110 pounds. A tank full of methanol would weigh about the same, but would give only about half the range.

Interest in methanol as an alternative fuel revived in the late 1990s for fuel cells. A number of strategists had argued decades earlier that hydrogen, normally a gas, might not be the preferred eco-fuel for automotive use after all and that perhaps some other liquid fuel might be preferable. Methanol first surfaced as a fuel as far back as 1974 at the first international hydrogen conference, the 1974 THEME conference in Miami. Researchers from the Stanford Research Institute said that replacing gasoline as a fuel was very difficult because the "distribution network proves to be a dominant component of the total private vehicle transportation system." "Once established," they noted, "the infrastructure networks of a system become very resistant to change. . . . Institutional change is often less readily accomplished than technical change." Societies get locked into a given system. The paper concluded that "compared to several alternatives, especially the use of methanol in vehicles, a transition to hydrogen would appear to be needlessly disruptive," something that is still argued in some quarters today.[9]

Methanol, a particular type of alcohol, was supported for many years as the preferred clean liquid fuel for cars, notably by Volkswagen beginning in the early 1970s, but also by DaimlerChrysler. Methanol is closest to hydrogen in terms of environmental cleanliness. It has only one carbon atom in its structure. It has been described as "two molecules of hydrogen gas made liquid by one molecule of carbon monoxide," and as thus sharing many of the virtues of pure hydrogen. Also called methyl alcohol, wood alcohol, or methylated spirits, methanol (CH_3OH) is a clear, odorless liquid that freezes at $-144°F$, boils at $148°F$, and mixes easily with water—one of the drawbacks when it comes to burning it in internal combustion engines

Methanol can be made from many sources, including natural gas, petroleum, coal, oil shale, limestone, and even wood, farm, and municipal wastes, an important consideration. Made from renewable plant matter, it would be a greenhouse-gas-neutral fuel, since it would not add to the world's total CO_2.

One of its chief drawbacks that in the past has precluded its wide use in internal combustion engines is its low energy content. Per unit volume, it has only a little more than half the energy content of gasoline—64,700 Btu per gallon versus 120,000 for gasoline: a methanol-fueled car gets only 55 to 60 percent of the mileage of a conventional internal combustion engine car. Another big drawback that had a lot of people worried was that methanol is toxic. A material safety data sheet published by Canada's Methanex Corporation, a global supplier of the chemical, said that "swallowing even small amounts of methanol may cause blindness or death" and that "effects of lower doses may be nausea, headache, abdominal pain, vomiting." A 1997 U.S. government monograph, *Methanol Toxicity*, warned on its cover page that "the shift to alternative motor fuels may significantly increase both acute and chronic methanol exposures in the general population."[10]

Today methanol has faded into oblivion as transportation fuel, but it is very much alive with vigorous development and intensive marketing as fuel for small and micro fuel cells for handheld electronic devices, portable fuel cells for off-grid and mobile applications, and communications and uninterruptible power supplies for the military, ranging from a few watts to several kilowatts. For example, at the Twelfth Small Fuel Cells 2010 conference held in April in Cambridge, Massachusetts, Peter Podesser, CEO of one of the leading, most successful manufacturers of small methanol fuel cells, Germany's SFC Smart Fuel Cell AG, said in his conference-opening presentation that his company has shipped more than 15,000 commercial fuel cell products in the past five years for mobile homes, sailboats, electric cars, and scooters; to power remote sensors; and, for the military, as field chargers or as lightweight power source for soldiers' electronics. (To be sure, the conference covered other small fuel cell systems as well, such as PEM, solid oxide, direct borohydrides, and hybrid configurations. Other presenters included Samsung Electronics, Nuvera Fuel Cells, and Neah Power Systems.)

For hydrogen, efficient onboard storage is still an ongoing concern in the U.S. Department of Energy's (DoE) hydrogen program, though less so among international car manufacturers. The concern prompted DoE to set up a "Grand Challenge" storage program in 2003, followed in 2008 by selecting ten organizations and laboratories as a virtual Hydrogen Storage Center of Excellence.[11] While major carmakers heavily involved with hydrogen and fuel cell development would welcome a storage breakthrough, they do not see it as absolutely necessary now and are content with compressed hydrogen onboard storage. Witness the Toyota fuel cell Highlander SUV, mentioned in chapter 1, which demonstrated a range of 431 miles.

One storage technique that researchers had begun to look at in the 1990s was storing hydrogen in carbon nanotubes. An early sensationalist report came from two Northeastern University scientists, Nelly Rodriguez and Terry Baker, who claimed in 1997 that they had developed a graphite nanofiber material that would store as much as 65 percent of hydrogen by weight—a mind-boggling number when most conventional storage systems reported single-digit percentages. The Rodriguez-Baker claims quickly faded into oblivion when nobody else was able to match their results. Two years later, four researchers at the National University of Singapore—P. Chen, X. Wu, J. Lin, and K. Tan—claimed to have achieved hydrogen storage of about 20 and 14 percent, respectively, with lithium- and potassium-doped carbon nanotubes at moderate (200–400°C) or even room temperatures and ambient pressures. Their finding was more modest than the off-the-wall Rodriguez-Baker figures but still much better than conventional methods.

In the United States and elsewhere, a great deal of work has been going on with nanomaterials for several years, albeit with far less fanfare and more modest storage capacity claims, at the National Renewable Energy Laboratory (NREL) and several other laboratories. At the spring 2000 DoE Hydrogen Program Review meeting, Michael Heben, head of the NREL effort, reported that his team had achieved storage capacities of up to 7 percent by weight at room temperatures and pressures in small samples of 1 to 2 milligrams, meeting or exceeding benchmark standards set by the DoE program. A problem was the availability of this esoteric material. "In 1995," said Heben, "there was less than 1 gram of single wall nanotubes in the entire world. Today, several labs produce a gram a day." In 2010, a "Nanotube Yellow

Pages" site (http://www.nanoten.com/ntyp.html) on the Internet lists twenty-one suppliers of nanotubes and nanostructured carbon, and prices have dropped from around $1,500 per gram as of 2000 to retail prices of around $50 per gram as of March 2010, according to a Wikipedia carbon nanotube site. A November 2007 article said that multiwall carbon nanotubes, which are easier to make, were being produced by the ton.[12] At about that time, Baytubes, a German manufacturer, opened a plant producing 30 metric tons per year, doubling its total capacity; it was shooting for 200 metric tons by 2009. But single-wall tubes were more complex and expensive to make. Idaho Space Materials of Boise claimed production rates of 50 grams per hour using a NASA-developed arc-discharge technology, and SouthWest NanoTechnologies of Norman, Oklahoma, said a new plant being built at that time would produce 1 kilogram of single-wall nanotubes per day at prices ranging from $50 per kilogram for high-purity commercial grades to $500 per gram for specialty grades. To be sure, other exotic-sounding storage technologies research has come into the picture since then. The 2009 DoE Hydrogen Program Merit Review included presentations on, to name just a few examples, aluminum hydride regeneration, catalyzed nanoframework stabilized high-density reversible storage systems, amine-borane-based chemical storage, hydrogen storage by spillover, nanostructured polymeric materials, and alternating metal-carbon layers.

As the new millennium was getting underway, fuel cells were receiving more and more attention from environmental groups and investors. A new wave of feverish fuel cell stock trading, similar to the dot-com frenzy that had roiled the markets previously, was cresting, followed by the almost inevitable rollback starting in the middle of the first decade of the new century in which hydrogen and fuel cell technologies faded from public perception, pushed aside for a while by biofuel, battery, and plug-in hybrid technologies for cars.

The Early Years

What follows is a rundown of some of the major milestones in hydrogen-fueled transportation during the past three decades, highlighted by two 1996 events: the unveiling of fuel cell vehicles by Daimler-Benz and

Toyota, and the unveiling of the first almost-close-to-commercial fuel cell passenger cars by major manufacturers.

In May 1996, three years before DaimlerChrysler's splashy spring 1999 showing of NECAR 4 in Washington and before Daimler-Benz merged with Chrysler, some 240 journalists from all over Europe and the United States attended a media bash in the middle of Berlin's Potsdamer Platz at which the wraps were taken off Daimler-Benz's NECAR 2 fuel cell minivan. Although the two-speed transmission hiccuped because of sensor and software troubles during demonstration rides, the event was a smashing success. The *Economist*, which reported the transmission trouble, said the new vehicle's "engine ran smoothly. And this was the important point, for the NECAR 2 is claimed to be the world's first car powered by fuel cells."[13]

NECAR 2 was the dramatically slimmed-down successor to the original, experimental 1994 NECAR 1, a large, boxy urban delivery van whose fuel cell took up just about all of the cargo space. NECAR 2 was a fuel cell prototype derived from Daimler-Benz's V-Class front-wheel-drive minivan. It was powered by two 25-kilowatt Ballard-type PEM fuel cells, one of the first payoffs of the initial $35 million cooperation and development agreement the two companies had signed in mid-1993. The white minivan, its raised roofline camouflaging compressed-gas tanks, ran on pure hydrogen. NECAR 2 had a range of at least 156 miles. Daimler-Benz engineers said that with prudent freeway driving, it could achieve up to 250 miles at speeds of up to 69 miles per hour between refueling stops.

In September 1997, the NECAR 3 follow-up was rolled out. It was an experimental fuel cell version of the pricey five-door A-Class subcompact, which Daimler-Benz was about to unveil with conventional gasoline and diesel engines in Europe. NECAR 3's 50-kilowatt PEM fuel cell was fueled by methanol, and the early rumor was that NECAR 3 might be the basis for the company's first production fuel cell vehicle. (It did not happen.)

In October 1996, Toyota unveiled its first fuel cell vehicle, the FCEV, a PEM-equipped version of the popular RAV4, at the Thirteenth International Electric Vehicle Symposium in October in Osaka, just about stealing the show from the almost two dozen electrics and hybrids presented by other carmakers. Toyota built two copies, at a cost of about

$1 million each. With the Toyota-developed fuel cell putting out 25 kilowatts, the FCEV had a specific power output of 0.12 kilowatts per kilogram and a range of 109 miles. In addition to power from the PEM fuel cell, it used a set of nickel metal hydride batteries to recapture energy otherwise lost in braking and deceleration. Whereas NECAR 2 relied on compressed hydrogen, the FCEV employed a titanium-based hydride. Its titanium-alloy tank stored about 2.4 percent hydrogen by weight. (The target was 3.2 percent.)

Toyota, like Daimler-Benz, was developing a parallel methanol-based version. "It's still too soon to know," Yoshio Kimura told a reporter when asked which version was likely to emerge as commercially viable. "No doubt we're moving to a hydrogen era, at which time we'd run this vehicle on hydrogen. But whether the infrastructure will be in place to coincide with our marketing plans remains to be seen. That's why we are developing both." Not to be outdone, Opel, General Motors's European subsidiary, rolled out a PEM fuel cell version of its then-new Zafira minivan in October 1998 at the Paris Auto Show. The van was powered by twin 25-kilowatt fuel cells, assisted by a 20-ampere-hour, 500-watts-per-kilogram, 6.3-kilowatt-hour Ovonic metal hydride battery pack. The fuel cell Zafira was the first product of GM's new Global Alternative Propulsion Center, an international fuel cell development facility with laboratories in Michigan, New York, and Germany.

The race was on:

- Ford unveiled a 75-kilowatt PEM-fuel-cell version of its lightweight P2000 research vehicle at the Detroit Auto Show in January 1999. The car, roughly the size and shape of the company's standard commercial midsize models but weighing about 40 percent less than its 3,400-pound standard cousin as a result of the judicious use of aluminum and other lightweight materials, was expected to perform just about like a Taurus, with 0 to 60 mph acceleration in 12 seconds or so. The car was not quite ready for road testing: "The January auto show came a little too soon for us," a Ford engineer in charge of the fuel cell project, Ron Sims, told a reporter. "We were in the final stages of integrating all the components which we had run successfully separately."[14]

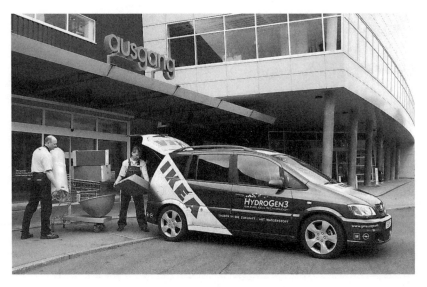

Figure 6.11
Two employees of Swedish furniture maker IKEA's Berlin store load a General Motors/Opel HydroGen3 fuel cell minivan for a delivery in the German capital in 2005. IKEA leased the liquid hydrogen–powered minivan as part of the Clean Energy Partnership Berlin cooperative fleet trial.

- At the 2000 Detroit show, Ford showed a fuel cell prototype called Th!nk FC5 (fuel cell, fifth generation). Based on the Focus model, it was equipped with Ballard's latest 75-kilowatt fuel cell stack, the Mark 900, designed to run on methanol.

- Also in early 2000, Korea's Hyundai announced a collaboration with International Fuel Cells to build two to four prototype fuel cell SUVs, the first ones perhaps in time for the 2001 Detroit Auto Show.

- In Europe, a joint effort by Renault and Peugeot was centered on a fuel cell under development by the Italian company De Nora. The Renault–Peugeot–De Nora program started out as a French effort partially funded by France's ADEME environmental agency. It widened in scope in early 1997 when it became the centerpiece of a new European Commission program dubbed HYDRO-GEN. The Daimler-Benz and Toyota premieres got wide media coverage, in large measure because they demonstrated the growing interest of major carmakers, in the past frequently derided as too conservative when it came to

Figure 6.12
A UPS driver explains the inner workings of his company's new fuel cell–powered 3-ton Dodge Sprinter in 2004 in Santa Monica, California, to a group of school children from LA's Best, an after-school program for inner-city students in the Los Angeles area. UPS operated four fuel cell vehicles in California and Michigan on UPS routes in a demonstration program with the Environmental Protection Agency, the DoE, and the state of California.

breaking environmentally benign ground. However, they were by no means the first. If we discount for the moment General Motors's experimental Electrovan and Karl Kordesch's Austin (see chapter 7), that distinction belonged to small, upstart companies that saw the fuel cell's potential much earlier than the giants.

• Three years before the Daimler-Benz and Toyota premieres, Energy Partners, a small company in West Palm Beach, Florida, unveiled a fuel cell vehicle that made a bit of a splash in the specialized media. Energy Partners's founder and chairman, the late John Perry (a millionaire and at various times a newspaper publisher, an operator of a cable TV system, and a builder of small submarines for oil exploration and for James Bond movies), had been tinkering with and supporting hydrogen, methanol, and fuel cell developments since the 1960s. Energy Partners's proof-of-concept car, a lightweight plastic-body two-seater made by Consulier Industries of West Palm Beach and converted to

PEM power by Energy Partners, took to the road in October 1993. Carrying three 15-kilowatt fuel cells in an open well, the car had a range of 60 miles in city driving and a top speed of 60 mph.

- A fuel cell–battery hybrid, a converted Ford Fiesta mail delivery car, had debuted in summer 1991 in Harrisburg, Pennsylvania. Pennsylvania's State Energy Office had chipped in about $60,000 in addition to what project developer Roger Billings said were millions of dollars from his own funds. Other support came from Air Products and Chemicals (Pennsylvania's preeminent producer of industrial gases) and from Exide (the largest U.S. manufacturer of lead-acid batteries). Exide had contributed the auxiliary batteries. The car's PEM fuel cell was rated at less than 10 kilowatts. The project faded quickly into oblivion.

Buses

As the new century began, early prototypes of both fuel cell and hydrogen internal combustion engine transit buses were being tested in North America and Europe:

- In 1991 the Belgian government gave formal approval to the Greenbus project. Put forward by the Belgian firm Hydrogen Systems, it called for the use of electrolytically produced hydrogen to power a converted 7.4-liter, 227-horsepower diesel city bus made by the Belgian bus maker Van Hool. The hydrogen was to be stored onboard in commercially available iron-titanium hydride storage vessels. The ZEMBUS (for "zero-emission bus") was launched in September 2000 in the town of Hasselt in the presence of Belgium's minister of transport.

- In April 1996 a bus running on liquid hydrogen was unveiled and put into regular daily operation in Erlangen, Germany. It had a 229-horsepower engine originally designed to burn natural gas. Another internal combustion–engine bus, operating on a mixture of natural gas and hydrogen called hythane, took to the streets of Montreal in Quebec in November 1995.

- In Brazil, where air pollution is choking the life out of urban centers such as São Paulo, hydrogen buses were given some consideration in the 1990s. Latin America's first hydrogen bus, a hybrid fuel cell vehicle, started road tests in São Paulo in August 2009, and three more were planned to be added later. Each was powered by two 65-kilowatt Ballard-designed fuel cells, and the buses themselves were built by Brazilian bus builders Marco Polo and Tuttotrasporti.[15]

- China issued a request for a proposal for a fuel cell–powered bus in early 1998. Six years later, it stunned Western participants in a Beijing hydrogen conference, HYFORUM 2004, organized by the German Future Energies Forum, by displaying a third-generation 100-kilowatt hybrid fuel cell transit bus built by Tsinghua University and powered by fuel cells designed by Shanghai ShenLi High Tech Co.[16]

- In 1987 the first fuel cell bus project in the United States began with a competition held by the DoE for the basic design of a hybrid bus to be used at Georgetown University. The competition called for a fuel

Figure 6.13
China's first small PEM fuel cell bus, converted to compressed hydrogen gas by Tsinghua University and Beijing Green Power Co. Its existence was reported by a Japanese researcher at the 2002 World Hydrogen Energy Conference in Montreal. The range of the twelve-passenger bus was said to be 165 kilometers (103 miles).

cell system plus a backup battery system to help with acceleration and climbing hills and to store energy recaptured during braking. Two teams were pitted against each other. The team headed by Booz-Allen Hamilton also included the Engelhard Corporation (supplier of the fuel cell) and the Chrysler Corporation. The other team, captained by the Energy Research Corporation (supplier of the fuel cell), included the Los Alamos National Laboratory and Bus Manufacturing Systems. Four years later, when the DoE declared the first group a winner, Booz-Allen Hamilton was no longer at the helm; the New Jersey–based H Power Corporation had been named the prime contractor. The DoE's charge to H Power was to come up with three 27- to 30-foot-long buses within thirty months. The buses were to be powered by phosphoric acid fuel cells fueled by methanol, which was to be reformed into hydrogen with an onboard reformer (a system judged at the time to be less challenging than pure hydrogen and PEM cells). The first Georgetown bus made its debut in Washington, D.C., in spring 1994 as part of Earth Day festivities. By that time, Engelhard had bowed out as supplier of the fuel cell system; a fuel cell manufactured by Fuji Electric was finally used. The bus itself was built by the Bus Manufacturing Corporation.

- Ballard's first proof-of-concept bus was rolled out in January 1993. All twenty-one seats of the small prototype were filled by team members and well-wishers when it took its first cautious trips on company grounds. That first bus could start on a 20 percent grade, maintain 30 mph on an 8 percent grade, and accelerate from 0 to 30 mph in 20 seconds, and it had a range of 94 miles. In August 1994, the same bus made its public debut carrying real passengers during the first "green" Commonwealth Games in Vancouver. Operating for about six hours every day for most of a week, it shuttled hundreds of spectators from a central dispatch point to the various athletic events.

- In February 1994, Ballard announced plans for a 40-foot, 60-passenger, 275-horsepower commercial bus with a range of 250 miles and, further down the road, a 75-footer. Development of a 60-foot bus was underwritten by various Canadian government agencies and the South Coast Air Quality Management District. Rolled out in 1995, the low-floor bus was built by New Flyer Industries of Winnipeg.

- In September 1995, the Chicago Transit Authority announced that it would test three Ballard-powered PEM-fuel-cell buses. The buses began carrying paying passengers through Chicago's Loop in autumn 1997. In March 1996, Ballard signed an agreement with Vancouver's transit authority for fleet testing of three buses. When the Chicago and Vancouver test programs ended in 2000, all concerned proclaimed their satisfaction with the results.

- Europe's first fuel cell bus, the Eureka, made its much-delayed debut in Brussels near the end of 1994. First announced in 1988, the Eureka was an articulated 59-foot eighty-passenger vehicle whose fuel cell and other components were housed in a two-wheeled trailer almost as long as the bus itself. A hybrid with an 87-kilowatt alkaline fuel cell made by the Belgian Elenco company plus nickel-cadmium (NiCad) batteries from French battery maker SAFT, it had electrical traction equipment from Italy's Ansaldo and a liquid-hydrogen fuel system contributed by Air Products of the Netherlands. The bus itself came from Belgian bus manufacturer Van Hool. In all, the partners and the member governments spent about $8 million to get the bus on the road before its demise a few months later after Elenco went bankrupt.

- The entry in the fuel cell bus sweepstakes that probably raised the greatest expectations was Daimler-Benz's prototype NEBUS (New Electric Bus), unveiled in summer 1997. Another product of the partnership between Ballard and Daimler-Benz, the 12-meter low-floor city bus was powered by a ten-stack, 250-kilowatt PEM fuel cell installed in the rear; 190 kilowatts were available for traction, electrical systems, and air conditioning. Gaseous hydrogen was carried in seven roof-mounted 150-liter, 300-bar gas bottles, giving a range of up to 156 miles—more than enough for average daily requirements. In spring 2000, DaimlerChrysler announced that a commercial version, the Citaro, would be available for delivery in two years (figure 6.14). In the sleeker production model, the fuel cell system is stashed on the roof to achieve a more balanced weight distribution and to extend the floor farther back. Close to three dozen copies eventually were put on roads in Europe, Australia, China, and Singapore, and a new, more efficient hybridized version was announced in 2009.

Figure 6.14
Carmaker Daimler's most recent version of its Mercedes-Benz Citaro fuel cell bus made its debut in the German port city of Hamburg in November 2009. This new Citaro FuelCELL Hybrid, is a hybrid version (previous ones operated on fuel cells only). Mercedes planned to build an initial small series of thirty buses, with the first ten slated to go to Hamburg.

• Also in spring 2000, two German manufacturers, MAN and Neoplan, rolled out two hydrogen-fueled PEM fuel cell buses at a Fuel Cell Day event in Munich. The 12-meter MAN vehicle, dubbed Bayernbus I, had been in the works since 1996 and was powered by a 120-kilowatt Siemens-KWU PEM fuel cell originally designed for submarine use.[17] Neoplan's Bayernbus II was a technologically more advanced vehicle with a unitized carbon-reinforced plastic body, regenerative braking, two hub-mounted electric motors, and an 80-kilowatt fuel cell developed by a small start-up company, Proton Motor, located south of Munich. In 2009, Proton unveiled the triple-hybrid fuel cell bus it developed with Czech trolley bus maker Skoda Electric and the Czech UJV Nuclear Research Institute, combining a 50-kilowatt PEM fuel cell for medium constant power for recharging, an advanced battery for medium constant speed, and an ultra-capacitor for high power demand during acceleration or uphill travel.

Figure 6.15
Fuel cell–powered Chinese sightseeing vehicles are lined up at a hydrogen fueling station supplied by American industrial gas supplier Air Products. Fifty of these were dispatched to shuttle athletes and government officials at the Asian Games and Asian Para Games in November and December 2010 in Guangzhou City.

The bus was scheduled to go into regular route service in Prague's metropolitan area.

Small Specialty Vehicles

The idea of using fuel cell variants of golf carts for shopping trips began in Palm Desert, California, a wealthy and environmentally aware resort community.[18] Hundreds of street-legal golf carts, equipped with turn signals and brake lights, are used there for shopping and commuting. Peter Lehman, director of the Schatz Energy Research Center at Humboldt State University, and Glenn Rambach, then an engineer at Lawrence Livermore National Laboratory, conceived the idea in the early 1990s, when the per-kilowatt cost of fuel cell power needed for cars was still thought to be way out of economic reach. Lehman and Rambach discovered there was a sizable market for small utility vehicles of this type—about 300,000 vehicles a year. (Like many other early ideas, fuel cells to propel neighborhood vehicles has not really caught on.)

The first hydrogen-powered golf cart, turned over to Palm Desert officials in August 1996, was basically a standard golf cart, retaining its original 2-horsepower motor but equipped with a PEM fuel cell in place of the battery. It consumed about 0.29 kilowatt hours worth of hydrogen per mile—roughly the equivalent of 125 miles per gallon of gasoline, according to a Humboldt State University release. This was followed by a Danish-built Kewet that Lehman had reconfigured to run on compressed hydrogen. More like a tiny car than a golf cart, the Kewet had a power output of 9 kilowatts (12.6 horsepower). With a range of about 30 miles, this neighborhood vehicle was turned over to SunLine Transit in 1999 as a pool vehicle and a small test bed for that agency's ambitious plans to convert its compressed natural gas (CNG) bus fleet to hydrogen power.

A bigger-is-better variant of stretch fuel cell golf carts was deployed at the Asian Games and Asian Para Games in November 2010 in Guang-zhou City, northwest of Hong Kong. More than fifty of the open-sided green-and-white eleven-seaters were used to shuttle athletes and govern-ment officials between sites during the games (figure 6.15). The vehicles' powertrains and chassis were designed and built by a consortium of Tongji University, the National Fuel Cell Vehicle and Powertrain Research and Engineering Center, and Shanghai Fuel Cell Vehicle Power Power-train, and the cars' bodies were manufactured by Nanjing NAC Special Purpose Vehicle Co., part of the SAIC group. They are powered by 4.8-kilowatt PEM fuel cells, have a range of 75 kilometers (47 miles) between refueling stops for the four-wheel independent drive version and up to 100 kilometers (62 miles) for a two-wheel central drive version. Top speed is 40 kilometers per hour (25 miles per hour).

Energy Partners also saw potential in hydrogen fuel cells for small utility vehicles. It developed two concepts, the Genesis and the Fuel Cell Gator. The Genesis, unveiled in February 1995 in Palm Springs, was powered by a 7.5 kilowatt PEM fuel cell and was said to be capable of carrying eight passengers or 2,500 pounds of cargo. It carried hydrogen and oxygen in pressurized containers. The Fuel Cell Gator, an experi-mental version of a light utility vehicle built by the tractor company John Deere, operated for a while at the Palm Springs airport.

A decade and a half later, British engineer and former race car driver and designer Hugo Spowers came up with a Kewet-like, but more sophisticated fuel cell two-seater for urban dwellers, the Riversimple (figure 6.16). Spowers made a splash with the quirkily cute hand-built car during its first outing in London in June 2009. Its 6-kilowatt fuel cell, built by a Singapore company, Horizon Fuel Cell Technologies, provides power to four wheel-mounted electric motors that recover about 50 percent of the energy. Collaborators include Oxford University and Cranfield University, and support came from the prominent German Piëch automotive dynasty long identified with carmaker Volkswagen and

Figure 6.16
Britain's Riversimple urban fuel cell car prototype, shown here during its public introduction in London in summer 2009.

sports and racing car producer Porsche (Sebastian Piëch is a member of Riversimple's core team) and from industrial gas supplier Linde Group via its British BOC subsidiary. Interestingly, Spowers does not propose to sell the cars but to lease them, covering maintenance, fuel, and other items. And taking a cue from the computer culture, he intends to publish the design as open source, enabling tinkerers and engineers everywhere to redesign the vehicle and even set up manufacturing operations of their own.

Internal Combustion, Liquid Hydrogen, Electrification, and the Long Haul

The two holdouts in the move to hydrogen fuel cell–powered transportation are BMW and Mazda. While just about every other major international carmaker has joined the rush to fuel cell technology as the chemical-fuel high road to automotive electrification, BMW and, to a perhaps less visible extent, the Japanese carmaker, a subsidiary of Ford, still see prospects for hydrogen internal combustion. engines: BMW with sophisticated electronic engine management and other refinements, and Mazda by steadily and consistently developing Wankel engines a type of internal combustion engine using a rotary design to convert pressure into a rotating motion instead of using reciprocating pistons. Mazda is the last major carmaker to build them. There are signs, though, that BMW may be rethinking the situation. In early 2010, on the twenty-fifth anniversary of its research and technology group, BMW broke new ground for the company with a uniquely configured research hybrid that combined a gasoline internal combustion engine driving the front wheels— another change for traditionally rear-wheel-driven BMWs—with a small 5-kilowatt auxiliary power unit (APU) PEM fuel cell from UTC Power in Windsor, Connecticut, which feeds a bank of supercapacitors that drive the rear wheels and also provides electricity for other onboard power systems. The basic idea is to provide range and speed for highway driving with gasoline, and zero-emission low-speed propulsion in the city with hydrogen and fuel cells, with both systems working in sync electronically.

Whether BMW will ever abandon internal combustion engines completely and switch to full fuel cell power is not known now; there were rumblings in 2010 that the company's top management was thinking hard about this. But at that time, BMW and Mazda were the lone survivors in this category, which a couple of decades earlier counted companies like Daimler-Benz, Mazda, and, notably, Japan's Musashi Institute of Technology in the forefront of developing hydrogen internal combustion engines. Starting in the late 1970s, BMW has built six generations of liquid hydrogen internal combustion–engined cars, with a fleet of Hydrogen 7s, the last version, ferrying VIP folks at the 2009 Copenhagen Climate Summit.

In summer 1996, an elegant charcoal-gray sedan with bold lettering on its side reading BMW Wasserstoff-Antrieb (BMW Hydrogen Propulsion) conveyed environment minister (and now German chancellor) Angela Merkel to the World Hydrogen Energy Conference in Stuttgart to give her keynote speech (figure 6.17). To general disappointment, Merkel said she liked hydrogen for the long term but did not hold out

Figure 6.17
German chancellor Angela Merkel did the honors during the cornerstone-laying festivities in April 2009 for what was described as the world's first, if small, hybrid power plant, capable of producing both electricity and hydrogen from renewable resources, in Prenzlau. With her are Brandenburg state officials, left, and executives of the Enertrag renewable energy utility.

much hope for it in the near term for economic reasons. Since then, she apparently has changed her mind. In 2009, during dedication ceremonies for what has been described as the world's first, small—5.5 megawatt-hour—hybrid power plant in Prenzlau in what used to be communist East Germany, Chancellor Merkel, a physicist by training, said she was "pleased" with "this future-oriented project," adding, "Integration of renewable energy and energy storage will be of critical importance for a secure and climate-friendly energy supply."

The early, undisputed leader in hydrogen for transportation was Daimler-Benz. Daimler-Benz started investigating hydrogen-powered vehicles in earnest in 1973, racking up half a million miles by the end of the last century. Its first effort was a hydrogen version of a minivan that was eventually introduced at the 1975 Frankfurt Auto Show. To carry the fuel, that van used a titanium-iron hydride, mainly because the Daimler-Benz researchers were in a hurry to demonstrate the general feasibility of hydride storage and hydrogen as an automobile fuel to a wide public. "We completed the engine in about two weeks," recalled Helmut Buchner, who ran the hydrogen-hydride program at the time.

Except for various test gauges mounted on the dashboard, that first Daimler-Benz hydrogen bus was indistinguishable from the production model. One had to look very hard to notice anything different on the outside. The gas filler cap was replaced by a small hydrogen intake valve mounted on the right side underneath the body, just behind the front wheel. And the exhaust felt moistly lukewarm, almost like the steam from a teakettle that is just warming up. This was the water exhaust vapor, which was partially recycled through the engine to prevent backfiring.

In the early 1970s, when I was a correspondent for *McGraw-Hill World News* and *Business Week* in Bonn, Daimler-Benz invited me to take a ride in that van around its Stuttgart test track. Nothing seemed out of the ordinary. Engine noise appeared to be normal, and the van went through the banked turns at a steady 60 miles per hour. At one point, the driver stopped the bus on a 10-degree incline and then drove off smoothly in first gear. The 17-gallon tank carrying 440 pounds of titanium-iron hydride was stored under a bench seat. To release the hydrogen, heat from the engine radiator was cycled via a heat exchanger through the hydride tank. Rather than a conventional carburetor, there was a *Gasmischer* (gas mixer) to blend air and hydrogen; the mix was then sucked through the manifold into the engine for normal combustion.

The largest early attempt to gauge the potential of automotive hydrogen technology was a four-year fleet test of ten hydrogen-powered Daimler-Benz vehicles in what was then West Berlin. Five dual-fuel (hydrogen and gasoline) station wagons and five full-size hydrogen-only vans were turned over, after intensive training, to drivers from a Berlin emergency medical service, the city government's car pool, an ambulance service, the German Red Cross, a study group for efficient energy utilization, the Berlin gas utility GASAG, and a Jewish community organization, according to a 1990 summary report.[19] Begun in May 1984, the test ended in March 1988, after racking up more than 160,000 miles. Another 238,000 miles were accumulated in Stuttgart by prototype vehicles before the start of the actual fleet tests.

The engines of all these vehicles operated in the "external mixture formation" mode, with hydrogen blown into the intake manifold.[20] This is technically relatively simple, requiring only low-pressure hydrogen gas released normally by hydrides. But it also results in fairly low power and irregular combustion—that is, backfiring. The solution used at the time—injecting water into the suction manifold in the upper load range—was an acceptable compromise in terms of fuel consumption and power output; it also cooled some critical components. The hydrogen was extracted from "town gas," or "coal gas," a mixture gases containing hydrogen (about 50 percent), carbon monoxide, methane, and volatile hydrocarbons using pressure-swing adsorption. Developed during the nineteenth and early twentieth centuries, it was widely used in both the United States and Great Britain until the widespread adoption of natural gas during the 1940s and 1950s.

There were some technical problems: reduced hydride storage capacity because of the hydrogen's relatively low purity, engine corrosion because of water injection, corrosion in the hydrogen piping system from road deicing salt, mechanical damage to heat exchangers, and exhaust due to lower road clearance caused by the hydride system weight. Still, this test was judged an overall success. "On the whole, experience was positive and this was in large measure due to the fact that the vehicles were well received by the operators and the maintenance and service personnel. It was possible to operate the vehicles practically all year round without limitations and their reliability was very satisfactory on the whole."[21]

BMW began to explore hydrogen technology in 1978. That year, its first liquid-hydrogen car was put together by Walter Peschka from parts and components provided by BMW. Peschka, a researcher at the German aerospace research agency DFVLR (later renamed DLR) in Stuttgart, and a colleague, Constantin Carpetis, devised the world's first liquid-hydrogen fuel tank specifically designed for passenger cars. A flattened stainless steel sphere with two separate shells (one inside the other, with a vacuum in between), it was capable of holding about 29 gallons of liquid hydrogen at a low operating pressure of 4.5 bars. Peschka and Carpetis also developed a semiautomatic liquid-hydrogen service station pump, which they said was so simple to operate that it might be suitable for a self-service station.

In June 1994, BMW displayed all four generations of its liquid hydrogen–fueled internal-combustion-engine-powered cars at a one-day engineering symposium to which it had invited some 300 journalists from all over Europe. The conference opened many eyes to the practicability and safety of liquid-hydrogen technology for cars. Wolfgang Strobl, BMW's principal hydrogen investigator at that time, showed dramatic slides illustrating liquid hydrogen's safety: Double-walled tanks filled with the super-cold fuel and having all safety valves blocked were cooked over high heat, violently shaken in long-term vibration tests, and rammed with a massive pole (simulating the impact of a highway crash). In a fire test, hydrogen began escaping from safety valves after sitting about ten minutes in an open fire; it burned without any other visible effect on the tank. The pole's impact produced a leak through which hydrogen escaped slowly, but, said Strobl, the tank did not explode

The third major carmaker to jump into the hydrogen pool was Mazda, which began to investigate hydrogen as an automotive fuel around 1986. That effort culminated in the futuristic HR-X (hydrogen rotary experimental) concept car unveiled at the 1991 Tokyo Motor Show—the first of eight hydrogen-powered Mazdas (two copies each of four different models). Visually and technically, the HR-X was the most radical of the lot. This hydrogen hybrid was powered by a 1-liter-displacement Wankel-type rotary engine mounted in the center of the chassis. The engine produced 100 horsepower running on hydrogen—outstanding performance for an engine of its size at the time. Mazda, the last remaining adherent of Wankel technology, said that the Wankel engine, with its separate intake and combustion chambers, was uniquely suited among internal combustion engines to burn hydrogen, being free of the kind of backfiring and ignition problems that plagued other hydrogen internal combustion engines.

The HR-X made its U.S. debut at the New York Auto Show in April 1992. A year later, Mazda introduced the successor HR-X2 at the Tokyo Motor Show, a much more conventional-looking car with a front-mounted rotary engine driving the front wheels and offering much better performance. More or less simultaneously, Mazda developed a hydrogen-powered version of its popular Miata sports car, the main difference being that the standard 1.6-liter four-cylinder engine was replaced by a 1.3-liter twin-rotor engine. The main rationale for this conversion project was that it provided a basis for comparison with three battery-powered Miatas that had been built earlier and were being operated by a Hiroshima utility. For sports car aficionados, the killing drawbacks would have been the 770 pounds added by the hydride storage system, the reduced performance, and the reduced driving range (only about half that of the standard Miata).

In 1994, Mazda converted two station wagons to hydrogen, again using the 1.3-liter twin-chamber rotary engine to evaluate durability, refueling characteristics, and general usability under real-world driving conditions over a longer term. The cars were used every day in the motor pool of Nippon Steel's Hirohata Works, which generates hydrogen as a by-product. Together these two cars covered 12,500 miles in a two-year test of mostly city driving.

Mazda made a couple of stabs at fuel cell power—one a hybrid fuel cell station wagon, the Demio FCV, unveiled in 1997 on the sidelines of the Kyoto Global Climate Conclave, and again in 2001 with a methanol fuel cell version of its Premacy minivan in 2001. But over the years, the company's focus has been the rotary engine, used in bifuel (hydrogen and gasoline) versions of its RX sports car, dubbed the RX-Hydrogen RE that was launched in 2004, and of the Premacy Hydrogen RE hybrid minivan, first announced in 2005. Both models have been leased in small numbers to customers, with the minivan going to government agencies and energy companies in Japan.

BMW's and Walter Peschka's involvement with hydrogen was an outgrowth of a project begun in 1979, in which Germany's DFVLR aerospace agency teamed up with the Los Alamos Scientific Laboratory (its name at the time; now it's the Los Alamos National Laboratory) and the New Mexico Energy Institute to convert a Buick Century sedan to liquid hydrogen power.[22] As Peschka recounted in his book *Liquid Hydrogen*, Los Alamos was in charge of the overall coordination, including management, engine conversion, tests, and methodology. DFVLR provided the aluminum liquid-hydrogen tank and an electronically controlled semiautomatic refueling station.[23] The car was modified by one of the early hydrogen pioneers, Roger Billings of the Billings Energy Corporation of Independence, Missouri. Peschka described the car's acceleration as "not completely satisfactory" at Los Alamos's altitude (7,300 feet). Though driven mostly around Los Alamos, it was also demonstrated elsewhere in the United States.

As early as 1973, scientists at Los Alamos had converted a sturdy half-ton Dodge pickup truck to liquid-hydrogen power. That truck had a spherical 190-liter aluminum storage dewar (a Thermos-bottle-like double-walled container in its bed, right up against the forward wall). The carburetor had been replaced by a gas mixer made by IMPCO Technologies, Inc., originally designed for liquefied petroleum gas and propane. Exhaust gas was recirculated to avoid uncontrolled preignition, knocking, and backfiring. The vessel's boil-off rate was less than 1 percent per day, but in actual practice—during filling and discharge of hydrogen, and during measurements—the rate was more like 3 percent. Since the vehicle was apparently a sort of enthusiasts' project with no real research funding and sponsorship, it underwent only limited testing, accumulating only about 300 miles over roads in and around Los Alamos.

Of the fifty or so hydrogen vehicles thought to have been built by the turn of the century, a sizable fraction were powered by liquid hydrogen. Peschka's book documented fifteen of them, including a Winnebago motor home and a Chevrolet Monte Carlo converted by the Billings Energy Corporation in the early 1970s; the four BMW generations; and, beginning in 1975, the first in a series of passenger cars built by the Musashi Institute of Technology.

The Musashi 9, a refrigerated truck, was shown in Yokohama at a New Energy conference in 1993. A conversion of a 4-ton tilt-cab Hino, it made ingenious use of liquid hydrogen's cryogenic properties not only to power the 6-liter, 160-horsepower diesel engine but also to keep produce, fish, or other perishables fresh during transport. It was visually distinguished from its conventional cousins by a 400-liter cylindrical liquid-hydrogen storage tank mounted vertically on the left side between the cab and the cargo box.

A great deal of the early hydrogen development work was performed by enthusiastic individuals and hydrogen advocates in the 1960s and the 1970s. Amateurs as well as professionals were beginning to construct cars and hardware, picking up where Rudolf Erren and his contemporaries had left off three decades earlier.

In 1966, for example—three years before Buckminster Fuller was telling college audiences how to use the world's energy "current account" instead of robbing the global energy "savings account"—a high school student piloted a wildly backfiring Model A Ford truck through the quiet residential streets of Provo, Utah. With the help of his father, a teacher, and some friends, sixteen-year-old Roger Billings had spent some three months and 800 working hours not only restoring the Model A but making technological history. That truck, which delivered little power but "an amazing repertoire of noises," according to Billings, was apparently the first internal combustion engine car in the United States fueled by hydrogen. Billings had started tinkering with hydrogen as a fuel two years earlier. "Those were the proudest days of my life," he recalled later, "driving that truck on a fuel they said would never work."

Interspersed between Billings's early effort and the Daimler-Benz prototype of 1975 was a great deal of research, much of it done by university scientists. "It is not the internal-combustion engine that pollutes our air, but its present fuels," declared the late Kurt Weil, then a professor emeritus at Stevens Institute of Technology, at the landmark 1972 Energy Conversion Conference in San Diego. The German-born Weil had experience with hydrogen going back to the 1930s, when he and Rudolf Erren suggested a scheme to use the excess capacity of Germany's electric power grid to produce hydrogen through electrolysis. Proposing a similar plan for the United States, Weil told his fellow researchers that the central element of such a system is the hydrogen internal combustion engine. Its multifuel and mixed-fuel version not only offers complete adaptation of all existing internal combustion engines; it also allows complete flexibility in phasing out hydrocarbon fuels; and reducing or eliminating air pollution from internal combustion engines. Perhaps most representative of the early hydrogen enthusiasts who were stimulated by the emerging environmental movement and wanted to exploit the unique properties of hydrogen, were four members of the Perris Smogless Automobile Association, a group of residents of Perris, a small California town south of Riverside. The four—a civil engineer, a newspaper publisher, and two aerospace engineers—shared a commitment to ridding the environment of what they considered its worst enemy: automotive pollution. Patrick Lee Underwood, one of the aerospace engineers, had worked for the aerospace firm Lockheed in various research programs. He had thought about hydrogen as an aviation fuel in the late 1960s after hearing about an experimental hydrogen-fueled B-57 bomber that the U.S. National Advisory Committee for Aeronautics had test-flown successfully in 1957. The *First Annual Report* of the Perris Smogless Automobile Association, published in 1971,[24] stated that the idea of "using hydrogen and oxygen in a standard reciprocating automobile engine had occurred to Mr. Underwood several years ago but no constructive effort had been expended; indeed, the idea was not very well developed, generally because it was thought to be under development as a natural course by the aerospace industry."

The actual project had gotten underway in December 1969, when Underwood had challenged an editorial, written by Dwight Minnich for his paper, the *Perris Progress*, in which Minnich had asserted that there was no solution to the emission problem of the internal combustion engine. A month later, Minnich, Underwood, and Fredric Nardecchia got together in Minnich's office to sketch out the outline of a hydrogen test program. A short while later, they were joined by Paul Dieges.

During the next year or so, the four tried to convert a 1950 Studebaker they bought for $10 from a junk dealer (the engine failed to restart after overrevving on too much hydrogen), a 1930 Model A Ford that ran well, and a newer Ford F250 pickup converted to liquid hydrogen and liquid oxygen. The group had received a grant from General Motors to enter the liquid hydrogen/liquid oxygen pickup truck in the cost-to-coast Clean Air Car Race sponsored by Caltech and MIT, but the project had to be abandoned because of various and managerial problems.

The most poignant and revealing part of the Perris group's "first annual report" (there was no second) was the balance sheet.[25] Expenditures totaled $8,038 for a whole year's work, with major individual contributions of $2,465.68 by Underwood, $1,828.68 from Minnich, and close to $400 each from Dieges and Nardecchia. General Motors chipped in $2,180 and the California Society of Professional Engineers $300. The biggest expenditures were $1,429.01 for patent fees, $1,142.46 in travel costs for the Clean Air Car Race, and $877.72 for exotic parts for the noncompeting Ford pickup truck. The hydrogen cost $241.84. The oxygen bill ran to $50.20.

The language of the report shows humility as well as a sense of mission and excitement. In the Preface to the report, the four principals described themselves as "average business and professional men with average means and the usual family and personal obligations." They continued: "We have, of necessity, pursued our regular livelihoods during the course of this effort. These remarkable results, which have vastly exceeded even our own expectations, have thus far been elicited by a commitment of only somewhat higher order than we might have otherwise directed toward civic, social, church or hobby activities, plus our vacations which all of us devoted to this project." The sense of outrage from these three engineers and a small-town newspaper publisher at the ravages on the environment is clear in the report's opening words:

Future historians, if there are any, may well record that the private automobile and not the nuclear bomb was the most disastrous invention of a society so obsessed with technology that it never recognized the failures of engineering run rampant until too late. For the automobile-freeway system is surely the most inefficient, dangerous, costly, and environmentally damaging transportation ever conceived; about the only real plus is flexibility, and the actual popularizing features are frighteningly Freudian. . . . survival demands that the air pollution caused by the automobile be eliminated almost immediately.

As time went on, other organizations became interested in the potential of hydrogen. An early example was a film clip produced by the Jet Propulsion Laboratory in the mid-1970s. First, the camera focused on the exhaust pipe of an idling car emitting white vapors. Next, a man held a drinking glass against the exhaust tube, allowing the vapors to cool and condense into a colorless liquid. In the startling final shot, the man lifted the glass to his lips and drank what was almost pure water—a stunt that in the 1980s and 1990s became an almost standard photo op for hydrogen. (In an early instance, Mayor Richard Daley of Chicago had to gulp the waters from an idling Ballard PEM fuel cell bus in September 1995 when the city started its fleet test; ABC-TV correspondent Ned Potter did the same in a *World News Tonight* feature on the promise of fuel cells in April 1997.)

Hydrogen power won one of its early public validations as well as public attention in the 1972 Urban Vehicle Design Competition. Billings, by then a graduate student in chemistry at Brigham Young University, entered and won the antipollution category with a hydrogen-fueled Volkswagen that far exceeded the existing federal clean air standards.

The runner-up in the antipollution category, and the winner of the overall design category, was a team of students from the University of California at Los Angeles who had entered a hydrogen-burning American Motors Gremlin. The captain of the UCLA team was a student named Frank Lynch. Right after the results were announced, Billings and Lynch decided to team up in the Billings Energy Corporation of Provo, Utah.

Around 1979, Frank Lynch formed his own company, Hydrogen Consultants, in Littleton, Colorado. Renamed Hydrogen Components in early 1997, the firm is now internationally recognized as one of the savviest constructors of equipment for hydrogen-powered vehicles. For several years in the early 1990s, Lynch was a member of the U.S. Department of Energy's Hydrogen Technology Advisory Panel, a committee charged with advising the secretary of energy. (He is now semiretired.)

During the 1960s and the 1970s, there were at least a dozen efforts by academic researchers or private companies to employ hydrogen, either alone or as a supplement to regular gasoline, in the quest for lower emissions—for example:

- In 1970, two University of Oklahoma scientists, Roger Schoeppel and Richard Murray, under contract to the U.S .Environmental Protection Agency, adapted a 3.5-horsepower four-stroke engine to run on injected hydrogen. The modifications included installing a second camshaft system to operate the hydrogen-injection valve and installing a new water cooling system to improve the cooling characteristics of the cast iron cylinders. Tests were encouraging.

- Two University of Miami researchers, Michael Swain and Robert Adt Jr., converted a Toyota station wagon, lent by a Miami Toyota dealer, to run on hydrogen. Swain and Adt devised a new fuel system, dubbed hydrogen induction technique, achieving roughly the equivalent of 42 miles per gallon of gasoline.

- Harold Sorensen of the International Materials Corporation described reforming gasoline into hydrogen at the 1972 San Diego energy conversion conference.

- In 1973, Siemens, the German maker of electrical equipment, announced a "crack carburetor" that was to use a catalytic process to break down gasoline into methane, hydrogen, and carbon monoxide. The project was dropped two years later because Siemens could not find any customers

- In 1974, the Jet Propulsion Laboratory proposed combusting hydrogen together with gasoline at "ultra-lean" conditions. Hydrogen was generated aboard the car from gasoline through the use of hot air (1,500–2,000°F). Only small amounts would be produced to extend the flammability limits of the fuel-air mixture downward, requiring less fuel with the same amount of air. The car did turn out to be more efficient and to produce less nitrogen oxide, but at a power loss and other emissions were still high.

- In Detroit in 1974 two General Motors engineers, R. F. Stebar and F. B. Parks, found that by injecting hydrogen into unleaded gasoline, the "lean limit"—the flammability limit of a fuel in—could be reduced further and nitrogen oxide emissions could be cut drastically, but at a cost: the engine produced about one-third less power, and hydrocarbon emissions went up dramatically.

Hydrogen work also was going on in what was then the Soviet Union. At the 1976 Miami Beach hydrogen conference, Soviet scientists said in hallway conversations that they had experimentally converted a couple of Soviet-built Fiats to hydrogen use. Two years later, a Moskvich sedan was also modified to run on gaseous hydrogen, according to a report in the March 1978 issue of *Eastwest Markets*, a business publication covering trade with the communist bloc. The story said that "the vehicle worked well on a test track and emitted pure water vapor as exhaust." A November 1978 article in the newspaper *Socialist Industry* said that Soviet scientists had successfully tested a Volga automobile on a mixture of gasoline and hydride-stored hydrogen.

In 1988, the World Hydrogen Energy Conference, held in Moscow, turned out to be an eye opener for many Western visitors. Soviet scientists presented seventy-five papers, about half of the total. Hydrogen R&D "appears to be a serious effort in the Soviet Union," observed Alexander Stuart, then president of Canada's Electrolyser Corporation and a key figure of the international hydrogen energy scene at the time. "It seems to enjoy a high level of attention."

Some Western scientists who visited research institutes as part of the program noted quite a bit of outdated equipment side by side with modern instruments. By and large, though, they came away impressed with the breadth and scope of the work, the thoroughness of research efforts, and the number of scientific workers in the field. One Soviet researcher estimated that some 500 scientists were active in hydrogen energy–related work. The roster of the conference listed about fifteen institutes spread across the Soviet Union.

One highlight of the conference was the keynote speech, in which the aircraft designer Alexei Tupolev detailed the efforts that had led to the maiden flight of a partially hydrogen-powered TU-155 commercial jet earlier that year. Another highlight was the liquid-hydrogen-powered RAF (Riga Automobile Factory) 2203 minivan displayed outside Moscow's International Trade Center on the last day of the conference. And there were papers and posters describing such projects as the conversion of a Lada (a 1970s Fiat design licensed to the Soviet Union) to hydrogen, the installation of a fuel cell in another RAF van, the conversion of several other RAF vans to gasoline-hydrogen operation, and plans for the installation of a 40-kilowatt hydrogen-air fuel cell in a Hungarian Ikarus bus.

The 1988 conference turned out to be the last hurrah for large-scale hydrogen work in the pre-collapse Soviet Union. Although Russian scientists continued to present papers on various hydrogen projects at international meetings, there was little evidence that any real efforts were still underway. Hydrogen R&D's center of gravity had shifted to the West.

7

Fuel Cells: Mr. Grove's Lovely Technology

It looked like a garden-variety, standard U.S. diesel-electric railroad freight locomotive, the kind that can be seen chugging and pulling mile-long freight trains across the country (figure 7.1). But the orange-and-black BNSF railyard switcher locomotive that was paraded in January 2010 before California governor Arnold Schwarzenegger at the railroad's East Los Angeles–area rail facility in Commerce was anything but that. An offshoot of transit bus fuel cell technology, the fuel cell–powered 127 ton railway switch locomotive is the biggest, most powerful land-based fuel cell hybrid vehicle yet. It was initially unveiled in August 2009 in Topeka, Kansas, in the BNSF (formerly Burlington Northern Santa Fe Railway) railroad maintenance terminal in Topeka, Kansas, the largest North American rail freight network that was bought by Warren Buffett in late 2009.

The locomotive's second outing in Commerce was a fairly low-profile event. Perhaps for competitive reasons, BNSF had been keeping fairly mum about this project; there was no press release on its Web site. Nevertheless, the railroad's CEO, Matt Rose, was quoted in a local newspaper story as saying, "We are very excited to take a leadership role in this emerging technology."[1] Schwarzenegger, explaining that this was the first hydrogen-powered locomotive, said, "Believe me, the world is watching this right now. They will be talking about it in Europe . . . they will be talking about it in China and Japan. . . . railroads and locomotives have a huge, huge comeback." This is a small locomotive to assemble trains, he added, but "eventually the technology will go also to the big locomotive. So this is the beginning stage, where it always begins."[2]

The fuel cell hybrid switch locomotive is the brainchild of Arnold Miller, CEO of Vehicle Projects in Golden, Colorado, who in 1994

Figure 7.1
A hybrid fuel cell–powered switchyard locomotive, converted to fuel cell power from a standard diesel-electric by Vehicle Projects, Denver, Colorado. It is shown here on its first outing in the summer of 2009 at BNSF Railway's maintenance yard in Topeka, Kansas, with vehicle projects engineer Kris Hess pointing out some of the features to visitors and media. BNSF operated it in tests at its Los Angeles area facilities. (Photo by Jamie Koerner)

founded the Joint Center for Fuel Cell Vehicles at the Colorado School of Mines, morphing two years later into the Fuelcell Propulsion Institute, which constructed a fuel cell mining locomotive and a fuel cell mine loader. Starting in 2004, Miller and his team modified a conventional diesel locomotive by replacing the engine with two 125-kilowatt Ballard Power Systems fuel cell stacks—the same type used in Citaro buses. The stacks are rated at 250 kilowatts of continuous power and more than 1 megawatt of transient power, fueled by 70 kilograms of 5,000 psi compressed hydrogen gas. Although that is enough pulling power for railyard operations, locomotives pulling entire freight trains need close to three times that power—about 3 megawatts.[3]

Thinking further ahead, Miller is now investigating a new, not to say unusual, concept: a supersonic tube vehicle (STV), that is, a fuel cell–powered train-plane hybrid "flying" inside a big tube filled with a hydrogen "atmosphere." A thumbnail description on Miller's Web site (www.arnoldrmiller.net) explains it in this way:

Because of high energy consumption and noise of supersonic flight, the speed of conventional air transport is limited by the speed of sound in air. The central concept of a new idea in high-speed transport is that operation of a vehicle in a hydrogen atmosphere would both increase sonic speed and dramatically decrease drag relative to air because of the low density of hydrogen. The supersonic tube vehicle, a cross between a train and an airplane, runs on a guideway within a hydrogen-filled tube or pipeline, is propelled by contra-rotating propfans, and levitates on hydrogen aerostatic gas bearings. Vehicle power is provided by onboard hydrogen-oxygen fuel cells. Hydrogen fuel is breathed from the tube.

Miller calculates the STV's maximum cruise speed could be 3,500 kilometers per hour (2,187 miles per hour); at 3,000 kilometers per hour (1,875 miles per hour), its energy consumption would be only about one-third of that of a Boeing 747, which cruises at 870 kilometers per hour (544 miles per hour).

BNSF's behemoth isn't the only example of fuel cell railroading. Three years earlier, in fall 2006, Japan's Railway Technical Research Institute had conducted the first operational test of a railway vehicle powered by a fuel cell. The institute had installed a 125-kilowatt Forza proton exchange membrane (PEM) fuel cell made by Nuvera, of Cambridge, Massachusetts, and Milan, Italy, in a 33-ton Kuya suburban commuter rail car. The rail car operated at speeds of up to 40 kilometers per hour (25 miles per hour), and the institute's announcement said then the successful tests brought fuel cells a step close to being introduced into railway operations.

At the other end of the scale, a University of Illinois team in 2009 built and successfully tested the world's smallest PEM fuel cell. The liliputian cell measures a mere 3 by 3 by 1 millimeter, delivers an average energy density of 254 watt-hours per liter and 1 milliampere current, and higher energy densities of up to 400 watt-hours per liter were thought likely.

The tiny cell, about the length and width of Abraham Lincoln's profile on a penny coin (figure 7.2), carries its own fuel—lithium aluminum hydride ($LiAlH_4$) takes up about 50 percent of the volume—and does not need power-hungry auxiliary balance-of-plant equipment such as pumps and valves. And while the tiny device is in fact a fuel cell that generates electricity from hydrogen stored in the hydride,

Figure 7.2
Dwarfed by Abraham Lincoln's image on an American penny, a miniature fuel cell, a PEM technology-based device measuring 3 × 3 × 1 millimeter, built and tested by a team at the University of Illinois at Urbana-Champaign and unveiled at the start of 2009 after two years of development. Delivering an average energy density of 2.543Wh/liter, it was claimed to be the world's smallest. Possible uses include powering bug-sized miniature aerial reconnaissance robots.

with a fuel utilization efficiency of about 80 percent, in practice it works much like a primary battery: once the fuel is exhausted, the entire tiny package is discarded or recycled, with no refueling or recharging of storage tanks, according to Saeed Moghadam, at the time a researcher and spokesman for the team. The system produces enough current for simple electronic systems such as long-range unattended surveillance sensors or for "cognitive anthropods"—insect-sized flying robots for the military in low-level reconnaissance. The developers claim it can be increased by about an order of magnitude for nonmilitary uses.[4]

Between these two prototypical extremes—the behemoth fuel cell locomotive and the tiny military PEM cell—fuel cells today come in all sizes, types, and variants: from huge megawatt-rated power plants such

Figure 7.3
A $99 pocket hydrogen fuel cell charger for small electronic devices, launched in mid-2010 by the Singapore firm Horizon Fuel Cell Technologies. It delivers 1.5 to 2.0 watt continuous power at 5 V/0.4A using a standard USB port and stores up to 12 Wh net energy, according to its manufacturer.

as the molten carbonate fuel cell (MCFC) plants built and sold by Connecticut-based FuelCell Energy, to 80- to 100-kilowatt hydrogen PEM fuel cells typical for most current prototype fuel cell cars and light-duty vehicles built by major carmakers; smaller and medium-sized systems for backup power, telecommunications, and, increasingly, forklift trucks and similar equipment, many of which run on methanol instead of hydrogen; and small systems for handheld electronic devices, such as a 1.5- to 2-watt $99 MiniPak pocket fuel cell charger launched in summer 2010 by Singapore-based Horizon (figure 7.3). It operates on hydrogen stored in a small hydride cylinder, about twice the diameter and a little longer than a D-size flashlight battery, that can be recharged at home (the differences of basic fuel cell technologies will be described later in this chapter).[5]

The Early Days: Dr. Kordesch's Austin and GM's Electrovan

Today a commercial fuel cell market is building internationally, albeit in fits and starts and at different paces in different parts of the world. A bird's-eye overview of the global situation, presented at the U.S. Department of Energy's (DoE) 2010 annual merit review and peer review, reported that in the largest fuel cell market for stationary, portable, and auxiliary power including forklifts, about 75,000 units have been shipped worldwide, with about 24,000 of them in 2009. Most of them came from companies based outside the United States. However, in terms of total megawatt power shipped, about 50 of the almost 120 megawatt-sized plants came from American companies. Several hundred test fuel cell vehicles were being driven in 2010 on roads worldwide, about 150 in the United States, and fuel cells have been installed experimentally in small boats, sailboats, large ships. and airplanes. both as principal power plants in motorgliders and as auxiliary power units in commercial jetliners.

The situation today is a long way from the 1960s and 1970s when a few pioneers installed the first experimental fuel cells in cars—a sign of painstaking, perhaps painful, progress. One of the earliest examples was a homely black import seen on Ohio's lakeshore roads in the 1970s. At first glance, the small sedan tootling around the Cleveland suburb of Lakewood seemed like just another foreign economy car. Only on closer look did its unusual features become apparent. Mounted conspicuously crosswise on the roof were six scuba-type gas tanks. In front of them, over the windshield, was a sign reading "compressed gas—flammable." This 1961 two-door Austin A-40 was the harbinger of a revolution that was not to burst onto the international scene until a quarter of a century later. Constructed in 1970 by an Austrian scientist, Karl Kordesch, who had been spirited out of his native country by the U.S. Army, it was the world's first practical fuel cell car.

Kordesch had left Vienna in 1953, ostensibly on a vacation with his family and his two small children. After a somewhat clandestine passage through the Soviet-controlled part of Lower Austria, he and his family eventually turned up in New Jersey with the U.S. Signal Corps under the auspices of Project Paper Clip, a U.S. government program to bring Nazi Germany scientists to the United States after World War II. Two years later he joined Union Carbide's research laboratory in Parma, Ohio, as a staff scientist working on batteries. At Union Carbide he eventually became a corporate research fellow; this meant, he recalled, "complete freedom, including worldwide travel, and getting more money without the need to become a manager." Kordesch produced many patents and published many scientific papers.

The air-hydrogen alkaline fuel cell of Kordesch's Austin produced a tepid 6 kilowatts in conjunction with a bank of ordinary 12-volt lead-acid batteries (for acceleration and hill climbing). The fuel cell's electrodes, donated by Union Carbide, had been used earlier in another ground-breaking fuel cell vehicle, General Motors's Electrovan, a six-passenger van that had been converted with Kordesch's help in 1967.

The Austin, which Kordesch acquired from a neighbor who had ruined the engine after driving it for four months without enough oil, was otherwise practically new. The fuel cell system, the batteries, the hydrogen tanks, and other auxiliary equipment made the car about 25 percent heavier, but its performance and range were adequate. The lightweight roof tanks, pressurized to 130 to 150 bars and holding up to 25 cubic meters of hydrogen, provided a range of about 190 miles. Refueling at a hydrogen-storage tank farm took only about 2 minutes. An electric motor with a peak power of 20 kilowatts gave a top speed of about 50 miles per hour—fast enough for the placid suburb of Lakewood and its environs.

Normally the carbon dioxide (CO_2) present in the atmosphere poisons the electrodes of alkaline fuel cells—one reason that transportation fuel cell developers concentrated on PEM-type fuel cells in the 1990s, since they are largely immune to the problem. Kordesch neatly solved that difficulty by installing a soda-lime air scrubber, which removed at least half of the air's CO_2 content of 0.03 percent. In addition, his cell's alkaline (KOH) electrolyte absorbed some CO_2, reducing contamination even more. (Kordesch said the electrolyte could easily be changed when it had taken up too much CO_2.)

Kordesch could drive his Electro-Austin immediately after turning the key, using the batteries for power until the fuel cell reached its optimum operating temperature of 60°C to 70°C within a few minutes. He increased the life expectancy of the fuel cell by shutting down the batteries between operating cycles, shutting off the hydrogen supply, and emptying the electrolyte into a reservoir; this exposed the hydrogen electrodes to air, effectively regenerating the catalyst and eliminating all parasitic currents.

Kordesch drove the little fuel cell car more than 13,000 miles in four years. In its practicality and usefulness, it was far ahead of its time, and by some criteria, it still is. It was genuinely revolutionary, although few people would have noticed, or known, at the time. John Appleby, director of the Center for Electrochemical Systems and Hydrogen Research at Texas A&M University and one of the world's foremost authorities on fuel cells, devoted almost three pages to the Kordesch car in his seminal *Fuel Cell Handbook.*[6] "The significant feature of this vehicle was its practicality," wrote Appleby and his coauthor, F. R. Foulkes. "It was essentially the project of an individual (though very knowledgeable) enthusiast, with access to the right materials. Kordesch produced a low-pollution vehicle with useful performance and range, with an easy-to-operate fail-safe system. The technology employed was quite old even in 1970, and present-day electrode construction would significantly improve vehicle performance and range."

Perhaps even more esoteric was an earlier Kordesch machine, a fuel cell–powered moped fueled by hydrazine (N_2H_4), a carbon-free nitrogen-hydrogen composite fuel best known as a propellant for small rockets used to position spacecraft or change their attitude with short bursts. Kordesch built it as a demonstrator for the U.S. Army, which was looking at fuel cells for use in artillery radio service. A photograph reproduced in Kordesch's 1996 textbook shows him in a helmet and a heavy parka, sitting on his moped on a midtown Manhattan sidewalk in front of the Union Carbide Corporation office building, with spectators kept at a safe distance by a velvet rope.[7]

Union Carbide had always insisted that the moped and the car were Kordesch's private projects, with the company providing support only. The reason for this was a fear of liability. Kordesch said when he went to New York City to demonstrate the moped on Manhattan's sidewalks, he had to take out insurance totaling $2 million for two days to cover possible injuries and accidents. Fortunately, nothing ever happened.

The moped appeared pretty normal except for a couple of metal boxes—the fuel cell and a nickel-cadmium (NiCad) battery—mounted in the frame ahead of and underneath the rider. Kordesch converted the Austrian-made Puch (bought at Sears, Roebuck) in 1966 and drove it on public roads during his years at Union Carbide. There were two 16-volt, 400-watt hydrazine-air alkaline fuel cells and a NiCad battery. The batteries could be switched in parallel or series for speed control. The range was about 60 miles on 2 liters of a 64 percent aqueous hydrazine, with an easy top speed of 25 miles per hour.

Kordesch was not the only person, and not the first, to try hydrazine for fuel cell power. In the 1960s, the now-defunct Allis-Chalmers Manufacturing Company built a small 3-kilowatt golf cart powered by that exotic fuel. Earlier, Monsanto Research Corporation had developed a 20-kilowatt hydrazine-air alkaline fuel cell system for a ¾-ton Army truck. In 1972, Shell Research in England put a 10-kilowatt hydrazine-air system in a Dutch-built DAF-44 car. In 1982, according to Kordesch's book, the Japanese Shin-Kobe Electric Machinery Company employed the technology for a small 3-kilowatt military power plant.[8]

Today it is difficult to pinpoint who should get the credit for the first practical fuel cell vehicle; however, a 20-horsepower tractor designed by W. Mitchell and demonstrated by Harry Ihrig of Allis-Chalmers in 1959 keeps cropping up in the literature as one of the first. It was propelled by a 750-volt, 15-kilowatt, 917-kilogram hydrogen-oxygen alkaline fuel cell, a spinoff from space technology employing a circulating KOH electrolyte, according to Kordesch. The tractor had just about all the characteristics of modern fuel cell technology: high-voltage bipolar stacks with porous metal electrodes, catalyzed with platinum, and using an asbestos matrix to immobilize the liquid KOH electrolyte.

The following other fuel cell projects got underway in the early years:

- Union Carbide achieved a significant innovation in the mid-1960s with the development of all-carbon electrodes for the U.S. Navy and also for an experimental stationary 90-kilowatt hydrogen-oxygen prototype fuel cell (really a battery, according to Kordesch) for the Ford Motor Company.
- The availability of thin composite carbon-metal electrodes prompted General Motors to investigate fuel cells for transportation. That was how Kordesch got into the act. In 1967 GM decided to convert one of its six-passenger Handivans to a UC alkaline fuel cell system. The Electrovan's 32-module fuel cell system developed a peak power of 160 kilowatts—more than three times the output of demonstration PEM fuel cells that Ballard and International Fuel Cells were developing in the late 1990s for automotive use, but also a lot heavier. The Electrovan weighed 3,400 kilograms, compared to about 1,500 kilograms total for the standard version. According to Appleby and Foulkes, 1,790 kilograms were directly due to the fuel cell and electric drive system, but somewhere between 450 and 680 kilograms were due to

overdesign and test instrumentation. Still, the fuel cell van performed about the same way that its standard cousin did, accelerating from 0 to 60 miles per hour in 30 seconds and having a top speed of about 70 miles per hour. The range was between 100 and 150 miles on liquid hydrogen and 44 to 60 miles with compressed gases. It was a magnificent effort despite a host of problems: excess weight and volume, a short lifetime of costly key fuel cell components (the fuel cell lasted only about 1,000 hours after activation), a lengthy and complicated start-up procedure, system complexity, problems with temperature control, and, perhaps most important, safety problems. The potential for leaks of liquid hydrogen and the liquid KOH electrolyte, as well as the use of both liquid hydrogen and liquid oxygen in close proximity to each other in one vehicle, are all just too dicey in case of a crash, dangers compounded by the system's high voltage (520 volts). Still, GM concluded, according to Appleby and Foulkes, that "the rate of progress in this field and the strong advantages of fuel cells are sufficient incentives to maintain this effort."

• In Sweden, ASEA constructed a 200-kilowatt submarine unit in the mid- to late 1980s; apparently some mishap prevented its completion.

• In Germany, the battery maker Varta and the electrical equipment builder Siemens assembled a demonstrator version of a fuel cell–powered boat with an electric motor.

• In England, the Thornton-Shell combine operated a truck equipped with a hydrocarbon converter that extracted hydrogen from fossil fuel and cleaned it up with a shift reformer for use in an alkaline fuel cell.

A History of Fuel Cells

"At Last, the Fuel Cell" was the headline of an upbeat three-page article in an October 1997 issue of the *Economist*. "A device that has been neglected for a century and a half is about to take its rightful place in industrial civilisation," said the article. The opening sentence read, "Lovely technology, shame about the cost: That is the usual comment on fuel cells—a method of generating power that is 40 years older than the petrol engine." An accompanying editorial, "The Third Age of Fuel," said, "Just as coal gave way to oil, oil may now give way to hydrogen."

In passing, the article mentioned that the principle of the fuel cell had been developed by an Englishman, William Grove, "a man who although he ended his career as a judge, began as a physicist."[9]

Grove (1811–1896) was a professor of experimental philosophy at the now-defunct London Royal Institution and a friend of the famous physicist and chemist Michael Faraday (1791–1867), who discovered electromagnetic induction, invented the dynamo, and did research on electrolysis. After experimenting with electrolysis, Grove reasoned that it should be possible to reverse the process and generate electricity by reacting hydrogen with oxygen. In a classic experiment first reported in 1839, Grove built what is considered to be the first fuel cell. Although its electricity output was rather small, Grove was encouraged. Three years later, he built a bank of fifty such cells employing dilute sulfuric acid as electrolyte. He also reported on a hydrogen-chlorine fuel cell and found that other carbon-bearing liquids—"other volatile bodies such as camphor, essential oils, ether and alcohol associated with oxygen"— "gave a continuous current."

Grove was quick to recognize the elegant symmetry of the processes of electrolysis and recombination of hydrogen and oxygen, but he soon abandoned work on a "gaseous voltaic battery" because his device was not able to produce enough power.

In fact, new information that surfaced about a decade ago indicated that a contemporary and friend of Grove, the Swiss scientist Christian Friedrich Schoenbein, first discovered the fuel cell effect and that both men should be regarded as originators of the technology. *The Birth of the Fuel Cell* by Ulf Bossel, founder of the European Fuel Cell Forum and cofounder of the German Solar Energy Society, argued that Schoenbein discovered the fuel cell effect but that Grove built the first fuel cell–type generator. Bossel's book includes what he says is the first-ever transcription and publication of the complete correspondence from 1839 to 1868 between Schoenbein and Grove, a project that Bossel said at the time took him ten years. In other words, both men should be regarded as "the parents of the fuel cell", according to Bossel.[10]

In the mid-1850s, attempts were made to develop a carbon-burning fuel cell. But nothing much happened until 1889, when, according to Appleby and Foulkes, a British scientist, Ludwig Mond, and his associate, Charles Langer, repeated Grove's earlier work. They were the first to call the device a fuel cell, and tried to make it more practical by replacing oxygen with air and hydrogen with impure industrial gas obtained from coal by the so-called Mond gas process. However, after achieving 1.5 watts at 50 percent efficiency, they dropped the project because of the high cost of the platinum catalyst, because the platinum was being poisoned by traces of carbon monoxide in the gas, and because of other problems. In 1896, an American engineer named J. J. Jacques constructed the largest system yet, which ran on coal and air at 82 percent efficiency, producing 1.5 kilowatts and (according to one account) operating for as long as six months at a time. Later, it was found that the efficiency was actually much lower because of a misreading of the chemical processes involved. Still, Appleby and Foulkes credit Jacques with being the first to think of a fuel cell as a device that could provide electricity for domestic use.

Francis Bacon's Contributions

The modern fuel cell technology that led directly to the machines used to provide electric power on the space shuttle began in 1932 with work done by Francis T. Bacon (1904–1992), an engineer associated with Cambridge University in England and a descendant of the renowned seventeenth-century philosopher-scientist Francis Bacon.

A practical-minded engineer, Bacon reasoned that the high cost of the platinum catalysts employed in systems of the type developed by Mond and Langer would never permit fuel cells to successfully enter the commercial marketplace. To get around this, Bacon decided to develop a hydrogen-oxygen cell with an alkaline electrolyte and relatively inexpensive nickel electrodes. The temperature had to be raised to slightly more than 200°C to make nickel sufficiently chemically active. This meant that the cell had to be pressurized to keep the aqueous alkaline electrolyte from boiling, which led to Bacon's discovery that pressurization made the cell more efficient. (At first, Bacon pressurized his fuel cell to about 220 bars; eventually much lower pressures were used.)

Working first with an engineering company (C. A. Parsons and Co.) and later at Cambridge University, Bacon built a single-cell unit in 1939. In 1946 he constructed an improved version of the single-cell design; it was followed by a six-cell device in 1954. In 1959, with help from the

British government through the recently established National Research Development Corporation, Bacon produced a forty-cell unit that delivered 6 kilowatts at 200°C and 38 bars pressure. In 1959, Bacon announced that he and his coworkers had built and demonstrated a practical 5-kilowatt unit with enough power to run a 2-ton capacity forklift truck as well as a welding machine. Two months later, Allis-Chalmers demonstrated the first fuel cell–powered vehicle, the 20-horsepower tractor mentioned earlier. In 1964, Allis-Chalmers built a 750-watt fuel cell system for the Electric Boat Division of General Dynamics to power a one-man underwater research vessel.

In the early 1960s, NASA discovered that fuel cells were suitable sources of electric power for space flights of up to fourteen days. Non-rechargeable batteries simply would not last long enough, and there was no place in space to plug in and recharge a NiCad battery. And fuel cells would not need sunlight as photovoltaic panels do. (In addition, photovoltaic panels require backup batteries on low-orbit flights, since the spacecraft is sometimes in the Earth's shadow.) Perhaps most important, a fuel cell system running on liquid hydrogen and liquid oxygen produced about eight times as much power per weight—1.6 compared to 0.2 kilowatt-hours per kilogram—as the best batteries then available. NASA awarded more than 200 contracts for research on the physics, kinetics, electrochemistry, and catalysis of fuel cell reactions; how to manufacture electrodes; and all sorts of other details.

General Electric fuel cells with ion-exchange membranes were first used on the *Gemini* orbital flights. The *Apollo* lunar missions used fuel cells based on Bacon's design and developed by the Pratt & Whitney Division of United Aircraft Corporation (later United Technologies), which had bought licenses for Bacon's fuel cell technology. In 1970, United Technologies won the contract to provide fuel cells for the space shuttle program, which lasted four decades and came to an end in 2010 when UTC Power, a United Technologies division, and its sister company Hamilton Sund-strand, shipped the last space shuttle fuel cell power plant after maintenance and overhaul to NASA's prime shuttle contractor, United Space Alliance.[11]

NASA's engagement laid the groundwork for the renewal of interest in fuel cell development that came in the 1990s. "The massive US aerospace fuel cell effort has undoubtedly provided the single most important impetus to the development of electrochemical engineering science in respect to energy conversion," wrote Appleby and Foulkes. "There were many predictions that [fuel cells] would be the solution to the world's energy problems," they added. A host of programs got underway in the United States, Europe (including what was then the Soviet Union), and Japan to design and build various types of fuel cells—molten carbonate, solid oxide, phosphoric acid, alkaline, solid polymer, and direct methanol—as primary electric power sources for utilities and portable and small-scale uses.[12]

Utilities and Fuel Cell Power Plants

In the late 1960s, the Edison Electric Institute, United Technologies, and a group of electric utilities began to investigate the expected advantages of fuel cell power plants, with help from the gas utilities represented by the American Gas Association (AGA). Beginning in 1967, the AGA funded a nine-year Team to Advance Research on Gas Energy Transformation program, with United Technologies as prime contractor and the Institute of Gas Technology as subcontractor. The goal was to develop small natural gas fuel cells for home use, with phosphoric acid systems as the main technology and molten carbonate as a backup.

The hopes began to fade when technical difficulties became apparent to the fuel cell community in the late 1960s and the early 1970s. Coupled to a parallel slowdown in aerospace programs, these obstacles almost led to the demise of fuel cell development for terrestrial applications. Appleby and Foulkes list four major problems:

- Hydrogen was the only useful nonexotic fuel, but using it with relatively inexpensive nickel catalysts in an alkaline fuel cell required high temperatures and pressures, costly pressure vessels, and ancillary equipment.

- Alkaline fuel cells required pure hydrogen. That was problematic when hydrogen was produced from common fuels such as natural gas or coal. Any residual CO_2 in the hydrogen reacts with the liquid alkaline electrolyte, gumming up the electrodes' microscopic pores and slowing the overall chemical reactions.

- The use of "dirty" commercial fuels plus CO_2-containing air—as opposed to pure hydrogen and pure oxygen used on spacecraft—made the useful life of fuel cell systems (using construction materials commercially available at the time) too short for economical operation.

- It became clear that the closely knit community of fuel cell designers, engineers, and scientists had "tended to oversell the merits of the fuel cell before really having come to terms with all the teething troubles of an immature technology," Appleby and Foulkes wrote. "As a result of overenthusiasm, deadlines were not met, and private funding for the fuel cell greatly declined. With the major budget reductions in the aerospace program during the early 1970s, government funding for fuel cells also slowed. As a consequence, fuel cell research and development almost came to a halt because the private sector was hesitant to assume the costs previously paid by the US government."

Interest in fuel cells on the part of many American electric utilities picked up again in the wake of the 1973–1974 oil embargo. The government resumed funding development work for large-scale stationary fuel cell power plants. Overall, Appleby and Foulkes estimated that the government spent about $350 million (in 1986 dollars) between 1977 and 1984 on the development of stationary fuel cells. Utilities and manufacturers contributed a roughly equal amount.

Big Fuel Cell Power Plants Appear: United Technologies' PC-19

Among the dozens of demonstration plants, subsystems, test facilities, and other related hardware constructed or planned in the United States, Europe, and Japan in those years, two examples stand out. The first example was the first megawatt-class fuel cell power plant: the PC-19 phosphoric acid plant built by United Technologies in South Windsor, Connecticut and tested during the first half of 1977.

In February 1977, as PC-19 began churning out electricity, the U.S. Department of Energy, in cooperation with the Electric Power Research Institute, issued a request for proposals to build a big (4.5 megawatts) phosphoric acid demonstrator fuel cell power plant in New York City. Consolidated Edison was picked as the host utility. The basic idea, Appleby and Foulkes said, was to demonstrate the feasibility of installing and operating a fuel cell power plant in an urban utility environment—as it turned out, a difficult task at the time: "From the viewpoint of obtaining an operating license, the site had to be considered as one of the most difficult in the United States." In general, Appleby and Foulkes wrote, the history of the Manhattan fuel cell plant illustrates "the need for patience associated with the development of an emerging technology under real-world conditions." The plant was to be built within a year on a three-quarter-acre site at East Fifteenth Street and Franklin D. Roosevelt Drive in Lower Manhattan. It was to start operating in 1978, and the test was to be completed in 1979.

The plant was built, but it never produced any electricity. In addition to the usual squabbles among contractors and subcontractors and late deliveries, the builders and the utility had to contend with the suspicion, fear, and ignorance of new technology on the part of New York's bureaucracy. For example, according to Appleby and Foulkes, the New York Fire Department, put in charge of overall public safety, decided initially that the plant was really a refinery and therefore would not be permitted in the city. The fire department later agreed to let it be built as an experimental power plant. Then a number of stringent and unorthodox safety tests resulted in irreparable damage to some one-of-a-kind advanced heat exchangers. To make matters worse, according to Karl Kordesch, water left in the heat exchangers during the winter caused the exchangers to crack.

In early 1982, three years after the tests were supposed to have been finished, the fuel reformer ran into trouble because of clogging; this was attributed to the fire department's insistence on burning high-flashpoint naphtha fuel for safety reasons, whereas the plant had been designed for low-flashpoint fuel. In March 1983 the plant extracted the first hydrogen from naphtha for testing. In May the plant ran for three minutes on regular process gas on one occasion; a month later, it ran for almost an hour. Verification of the transition from standby to load completed the crucial process-and-control test.

In the spring of 1984, all the fuel cell stacks were finally installed. It was then discovered that they did not function properly. The stacks, which had been in storage for seven years, were found to have had a limited shelf life because of the chemical interaction of the small amount of phosphoric acid electrolyte and the graphite bipolar plates between the cells. Design changes were required to prevent this from recurring. (Kordesch says they had dried out, something that could have been prevented by the addition of some extra phosphoric acid.) Ordering new stacks would have added another $11 million to the plant's cost, plus another year and a half, to a project whose construction cost had already doubled from the original $35 million. At that point, the partners threw in the towel. "The only logical step," according to Appleby and Foulkes, "was to terminate the project and clear the site, which was done by December 1985."

Although the Manhattan plant never produced any power, Appleby and Foulkes argue that the effort was not a waste and that it provided valuable lessons. First, it had passed all the other tests required to confirm the system viability. Most important, and in spite of the responsibilities and complexities involved in the codes at a major urban site, the plant showed that it was possible for a fuel cell power-generating station to comply with the demands of the local authorities. Second, after extensive extra testing under nondesign conditions, which resulted in considerable damage and long delays, it received its license to operate.

Tokyo Electric Power Company Signs Up for Fuel Cell Plant
In 1980, the Tokyo Electric Power Company (TEPCO), the largest privately owned electric utility in the world, signed a contract with United Technologies for a 4.5-megawatt plant of the same type as the New York plant to be built in Goi in the Chiba Prefecture. In Japan, the construction and licensing procedures were simple, although special measures had to be taken to guard against possible earthquake activities, high tides, and oil spills. The TEPCO version incorporated a number of technical improvements, in part results of the New York experience. The first tests at half load (2 megawatts) started in April 1983, and the plant began operating at full power in February 1984, producing power at rates that exceeded design specifications. To be sure, there were some problems, such as the fact that start-up took 6 to 8 hours rather than the planned 4 hours because of some incorrectly dimensioned plumbing. The plant operated successfully for two years on reformed methane, the longest run (500 hours) demonstrating dramatically improved reliability, according to Appleby and Foulkes. No deterioration in performance was seen in 2,800 hours of operation.

The first Goi plant had been built within its budget of about $25 million—about 60 percent of the cost of the New York plant—and only thirty-eight months after the order was signed. "Considering the problems that had occurred with similar subsystems in New York," Appleby and Foulkes wrote, "completion of the experiment in Goi must be regarded as a remarkable achievement, which leaves in no doubt the possibilities of properly designed fuel cells operating in a utility context."

Next, the International Fuel Cell Corporation began to develop an 11-megawatt plant for TEPCO, in cooperation with Toshiba Corporation, at Goi. Site preparation began in January 1989, and installation of the first of three groups of six stacks each, with 469 cells per stack, began in June 1990. Plant start-up occurred on March 7, 1991, and it achieved an output of 11 megawatts before the end of April. By August, the plant had produced 10,264 megawatt-hours of electric power in 1,414 hours of operation, including one uninterrupted stretch of 875 hours. Initial evaluations found that the measured efficiency exceeded the design value and that the stacks were in "extremely good condition," according to an interim report.[13] Some of the fuel cells leaked as a result of an electrochemical reaction between the reformer exhaust gas and the carbon electrode material, some of the leaks being "dozens of times greater" than the values recorded in acceptance inspections, but others showed almost no leakage. Several other breakdowns occurred in this phase (as is normal when a new technology is being developed), but none of them serious. The plant, intended only as a technology demonstrator, operated until March 1997, producing more than 77,000 megawatt-hours in more than 23,000 hours of operation.[14]

Around 1993, United Technologies, through its former International Fuel Cells Corporation and ONSI subsidiaries, began building 200-kilowatt phosphoric acid fuel cell plants. More than 270 of these plants have been deployed around the world at a rate of about two dozen a year since the program started, the only demonstrator fuel cell plants at the time deployed in large numbers. Since they were demonstrators only, most of them have now been retired. In mid-2010, 65 plants were still operating in the United States and 35 abroad, according to UTC Power's media office. A year earlier, UTC Power had begun delivering the first of its new 400-kilowatt stationary fuel cell plants with a ten-year stack life and twenty-year overall system life, twice that of the previous model; the company planned to deliver fifty of them in 2010 and seventy-five in 2011. Overall, the number of fuel cell power plants of various types around the world is still small: A Worldwide Stationary Fuel Cell Installation Database on the nonprofit Fuel Cell 2000 Web site (www.fuelcells

.org) lists twenty-seven existing or planned installations of 1 megawatt or more (most of them in the United States, six in Korea, a couple in Japan, and one in Germany), and ninety plants in the range of 250 kilowatts to 1 megawatt, most of them in the United States (more than fifty), thirteen in Japan, seven in Germany, three in Italy, two in Korea, and others in Turkey, Italy, Switzerland, Canada, Singapore, and the Slovak Republic.

As Appleby and Foulkes observed, it was once widely assumed, on the basis of the early space-related efforts, that small fuel cells (those putting out a few kilowatts) for electric vehicles would come first, and that larger units for stationary power generation would follow. However, in the late 1980s, it seemed that large, multi-megawatt stationary systems would come into commercial use first. "Only a large assured market can justify the major public and private expenditures necessary to ensure a proper developmental effort," Appleby and Foulkes argued. Another reason, they say, may be that big stationary fuel cells do not face the resistance of strong lobbying groups (such as automobile manufacturers, which "might prefer to use the internal-combustion engine for as long as is profitable").

Fuel Cells for the Home and for Cell Phones

Efforts to commercialize home-size fuel cells—in part an outgrowth of the deregulation drive in the electric utility industry in the mid-1990s—resumed in the second half of the 1990s. These units were to be capable of operating on readily available natural gas and also on propane or other fuels, which could be delivered to islands and other remote areas. A California-based group organized by U.S. and Canadian utilities, the Small Scale Fuel Cell Group, was apparently the first to issue a market opportunity notice for such systems in 1997, essentially asking for bids and guaranteeing purchase of a dozen or so units for demonstration purposes and eventual mass production. A March 2010 overview piece on residential fuel cell trends by Fuel Cells 2000 staffer Sandra Curtin reported that more than 3,500 residential-scale fuel cells have been deployed or demonstrated worldwide in backup power, uninterrupted power supply or residential applications, with most of them sponsored by government agencies. Because of higher electricity costs in Japan and Europe, residential fuel cells are expected to be competitive with existing home energy technologies much sooner than in the United States: Japan has already installed thousands of PEM home units with government assistance, and the country's ultimate goal is for "two million homes to be powered by fuel cells by 2020," Curtin wrote. Similarly, Germany's "callux" field test program planned to install about 800 PEM and solid oxide fuel cells in homes by the end of 2012, but because of technical problems, the installation rate was behind the initial target, and only 111 units had been installed by January 2011, according to a report presented March 1 at a workshop in Tokyo organized by the International Partnership for the Hydrogen Economy (IPHE).

Even further down the scale are tiny fuel cells that their developers say will power cellular telephones, laptop computers, and electric hand tools. One of the first to surface was a fuel cell developed by a former Los Alamos weapons specialist, Robert Hockaday, which he claimed to be capable of giving about fifty times more talk time when used in a cellular phone than conventional NiCad batteries at what Hockaday hoped would be half the cost. Manufactured using lithographic etching techniques of the kind used in the manufacture of computer chips, it would be refueled with a squirt—about an ounce and a half—of methanol. The device, announced near the end of 1997, was initially designed to produce only enough power (about 4 watts) for a cell phone, but Hockaday expected to scale it up to produce the roughly 30 watts required by a laptop computer. In 2009, Hockaday and his company, eQsolaris, were still plugging along but had shifted the focus from small fuel cells to photovoltaics, working on micro photovoltaic concentrator arrays, according to a January 2010 article in the *Los Alamos Monitor*.[15]

After years of hype and hollow promises, small handheld fuel cells were beginning to sidle into the marketplace by the end of the new century's first decade. Among the earliest ones to appear to considerable international hoopla were devices from Medis Technologies, headquartered in New York but operating out of Israel. Medis had offered various devices, including a handheld power generator based on liquid borohydride technology that had to be squeezed to make it go. It all ended in November 2009 with the resignation of two key executives, appointment of a Tel Aviv receiver, and delisting of the company from NASDAQ.

A more credible launch came in the fall of 2009 when electronics giant Toshiba announced its small lightweight Dynario direct methanol fuel cell, produced in only 3,000 copies and available only in Japan through Toshiba's online store. The 29,800 yen ($326) device was claimed to hold enough fuel to triple the battery life of mobile phones, music players, and similar devices via a USB cable.[16]

Another contender came in June 2010 with a $99 hydrogen-fueled MiniPak charger set (two fuel cartridges, tips for main cell phones, micro and mini USB adapters) by Singapore-based Horizon Fuel Cell Technologies, which had made a name for itself with a series of toy products, including a small fuel cell racer named by *Time* magazine as a Best Invention in 2006, fuel cell bikes, the fuel cell for the Riversimple city car (chapter 6), and the power plant for a small model jet plane, dubbed Hyfish, that the company had developed jointly in 2007 with the German aerospace agency DLR and the Swiss firm that designed the plane, Smartfish. Horizon expects the price eventually to drop to $29, and the rechargeable Hydrostik hydride cartridges priced at $9.99 initially are expected to cost a little more than half of that once volume production and sales take off. Horizon is also offering a desktop hydrofill refueling station as well as a foldable solar power accessory for home refueling. Each Hydrostik charge holds about the same amount of energy as seven to ten AA batteries, depending on use; refilling costs are said to be mere pennies.

Beyond the horizon are more exotic devices, such as the butane-fired solid oxide micro fuel cell under development by Lilliputian Systems in Wilmington, Massachusetts. The company says that consumer portable electronics are the target market with something the company calls a fuel cell chip manufactured with microelectromechanical systems-based (MEMS) fabrication techniques, but beyond that, the company has kept mostly quiet. Nevertheless, it has talked to some reporters, and a February 2010 article in boston.com, an online offshoot of the *Boston Globe,* said a major challenge has been to find ways to contain in such a small volume the high temperatures of up to 1,500°C produced by solid oxide fuel cells.[17] Apparently it had succeeded: In mid-2010, Lilliputian said on its Web site that it has built a small-scale pilot manufacturing facility and embarked on developing its initial prototype products, has completed development of its product prototype on schedule, and is testing devices and signing commercialization agreements with select customers.

As to micro fuel cell production volumes, the estimates run all over the place. One projection slide presented at the 2010 DoE Annual Merit

Hydrogen Program Review said that about 24,000 fuel cells were shipped worldwide in 2009, including somewhere between 8,000 and 9,000 portables—primarily toys. Other market researchers have cited numbers in the hundreds of thousands, which have been interpreted as fuel cells for toys but those estimates may have been on the high side as well.

How Fuel Cells Work

Fuel cells are directly related to conventional primary batteries (in which a zinc casing reacts with manganese dioxide to produce electricity) and to rechargeable secondary batteries (which use lead and lead dioxide or nickel and cadmium to store electricity). But whereas a primary battery simply runs down after its zinc casing is exhausted and a secondary battery must be periodically recharged, a fuel cell keeps running as long as it is supplied with fuel (e.g., hydrogen), just as an internal combustion engine or a gas turbine will run as long as it is fueled.

A single fuel cell is essentially an electrochemical sandwich a fraction of an inch thick, with a negatively charged anode on one side, a positively charged cathode on the other, and an electrolyte (a watery acidic or alkaline solution or a solid plastic membrane that permits the migration of electrically charged hydrogen atoms from the anode to the cathode) in the middle. Many individual cells can be stacked to produce a usable amount of electricity.

Fuel cells operate almost silently, generating low-voltage direct current from the catalytically aided electrochemical reaction between a fuel (hydrogen or a hydrogen-rich carrier such as natural gas) and an oxidizer (oxygen, either taken from the ambient air or carried in separate storage vessels).[18] Whatever sound is associated with them comes from auxiliary equipment: pumps to transport the fuel through the system and remove the water produced in the reaction, a blower or compressor to cool the stack and to convey oxygen to the fuel cell, and perhaps a humidifier (required by some types, such as PEM cells).

Practical fuel cell systems are complex to design and build, especially small, rugged ones for cars, trucks, and buses, which must stand up to bumps and temperature variations. (This is one reason it took so long to put the first prototypes on the road.) But the basic principle of how a fuel cell works is fairly straightforward.

In a hydrogen-oxygen fuel cell with an acid electrolyte, molecular gaseous hydrogen is fed in on the anode side. There electrons are stripped out of the H_2 molecules, leaving positively charged hydrogen ions. These migrate through the electrolyte layer to the positively charged cathode, where they combine with oxygen to form water. Meanwhile, the liberated electrons (electric current) flow from the anode through a wire or some other metallic material back to the cathode, performing work—for example, operating an electric motor, lighting up a light bulb, or powering a cell phone.

Materials other than hydrogen can serve as ionic energy carriers too—for example, carbon (as a carbonate ion) and oxides of various elements. "In theory, any substance capable of chemical oxidation that can be supplied continuously (as a fluid) can be burned as the fuel at the anode of a fuel cell," Appleby and Foulkes explain. "This overall reaction may be viewed as the cold combustion of hydrogen with oxygen"—cold combustion because it takes place at temperatures much lower than those in a conventional open-flame process. In normal combustion, all the energy generated is released as heat. In a fuel cell, however, part of the free energy of this electrochemical reaction is released directly as electricity; only the remainder is released as heat—obviously a much more efficient process for powering electric equipment and electric cars.

An important aspect that has assumed great importance in recent decades because of the concerns over global warming is that the preponderant nitrogen component in our atmosphere—about 80 percent of the air we breathe is nitrogen—plays no part in cold combustion. Only hydrogen and oxygen react with each other in a fuel cell. By definition, this process precludes the production of nitrogen oxides, a key component in air pollution. In contrast, any open-flame process (such as the combustion of gasoline in conventional car, bus, and truck engines and jet turbines) produces these harmful emissions.

There are several basic types of fuel cells, differentiated by their electrolytes and the temperature ranges in which they operate.[19]

Alkaline Fuel Cells

This is the type of fuel cell that Kordesch and other early pioneers favored. Beginning in the mid-1960s, a United Technologies division, now known as UTC Power, built and serviced them as power sources for the *Apollo* space missions and the space shuttle orbiter, right up to 2010 when NASA announced the end of the shuttle program. In their final, most refined space shuttle version, alkaline fuel cells have evolved to levels of unparalleled efficiency and low weight, operating at 60°C to 90°C and efficiencies of 50 to 60 percent in converting chemical fuel to electricity. The electrolyte is 35 to 50 percent KOH. While the principal applications have been in space, their advocates still argue they are the cheapest fuel cell technology, in part because they do not require platinum as a catalyst. The main disadvantage of alkaline fuel cells (AFCs) is that ambient CO_2 can degrade their performance. However, as Kordesch demonstrated with his Austin, there are simple and economical industrial technologies for purging CO_2 from alkaline systems.

Today, few companies and institutions are working this technology, and AFCs have not received a great deal of attention. One exception is the U.K.'s AFC Energy, which thinks that alkaline fuel cells are what clean energy advocates want—making electricity via underground coal gasification. It has teamed with Australian clean coal developer Linc Energy, testing the idea in 2010 by successfully producing clean electricity from hydrogen-rich synthetic gas fed to an AFC alkaline fuel cell at Linc's Chinchilla demonstration facility in Queensland. A June announcement by Linc quoted its CEO, Peter Bond, as hailing the test as a "major innovation." Bond said the trial run showed that "this is a feasible route to achieve the ultimate in clean electricity from stranded, sub-economic coal, of which there is an abundance in the world."[20]

Proton Exchange Membrane
PEM fuel cells operate between 50°C and 80°C.[21] Their efficiencies range from 50 to 60 percent (the latter claimed, tank-to-wheel, in 2008 by Honda for its Clarity car). They use polymer-type membranes of the Nafion type (produced for more than twenty years by DuPont and more recently also produced by Ballard, W. L. Gore, Asahi Chemicals, and others). More advanced high-temperature membranes that would deliver higher performance are under development. As made abundantly clear earlier in this and the preceding chapter, major international carmakers are pinning their hopes on these—and spending buckets of money—in the race to develop a viable fuel cell engine for cars and buses. PEM-type systems are the fuel cells of choice for transportation because of their

fast start-up (they can produce usable amounts of power almost instantly, even when close to the freezing point of water), high power density, and relative ruggedness.

Beginning around the middle of the 2000 decade, a high-temperature variant of PEM technology began to show up in research laboratories around the globe. High-temperature PEM fuel cells could benefit from faster electrode kinetics, a simplified system design, and, importantly, higher tolerance toward impurities in the hydrogen fuel, according to the abstract of the chapter on this technology in *Hydrogen and Fuel Cells,* a book given to all participants at the 2010 Eighteenth World Hydrogen Energy Conference in Essen, Germany.[22] High-temperature PEM fuel cells can theoretically operate at temperatures as high as 220°C, said the chapter's author, Christoph Wannek, a research specialist with the Fuel

Figure 7.4
China's president Hu Jintao is briefed on PEM fuel cell technology during a 2005 visit to Ballard Power Systems in Vancouver by Ballard's CEO at the time, Dennis Campbell. The visit came during Hu's ten-day visit to Canada. Media reports at the time said Hu, a hydraulic engineer by training, expressly asked to see Ballard, the only company he visited.

Cell Research Institute at the Juelich Research Center in Germany, but about 180°C is the practical upper limit so far. Wannek described BASF Fuel Cells as the biggest player in this field; the company launched the new technology in 2004. Carmaker Volkswagen announced research on a high-temperature PEM cell of its own near the end of 2006. Since then, several other companies and research groups—Japan's Samsung, the U.S.'s Plug Power, and Denmark's Serenergy, for example—have begun to work on variants of these devices.

Phosphoric Acid
Typically designed for stationary power applications, phosphoric acid fuel cells (PAFCs) operate with about 55 percent efficiency at temperatures of 160°C to 220°C, up to 90 percent in combined heat and power applications when waste heat is used for cogeneration. The electrolyte consists of concentrated phosphoric acid. The principal manufacturer is United Technologies' UTC Power division, known previously as UTC Fuel Cells and International Fuel Cells, which had installed more than 270 demonstrator units since the early 1990s, of which about 101 were still operating in mid-2010. Until the late 1990s, the PAFC was the only type that was commercially viable, but since then, other types, such as molten carbonate, have entered the stationary power field. For a few years, the sale of International Fuel Cell's PAFCs had been supported by the U.S. government with a small subsidy program as a sort of icebreaker to induce utilities and other commercial users to accelerate the phasing in of fuel cells of all types in the years ahead. The fuel of choice is natural gas.

Molten Carbonate
Molten carbonate fuel cells, which run at 620°C to 660°C, have been gaining ground in the recent years, much of it due to the efforts of Connecticut-based FuelCell Energy and, importantly, its Korean partner POSCO Power., as well as its former German licensee MTU (the license expired in December 2009). FuelCell Energy says some sixty of its trademarked Direct FuelCells are supplying baseload power in five countries. Their efficiency is typically in the 60 to 65 percent range and can be as high as 85 percent with waste heat utilization and cogeneration. As the name implies, the electrolyte is a melt (a binary alkaline carbonate

mixture), a corrosive material that has presented degradation issues. Energy omnivores, molten carbonate fuel cells' big attraction is that because of their high operating temperature, they can use and reform all sorts of fossil-based fuels—propane, gases from food processing and wastewater treatment plants, but also diesel fuel and coal gas—internally into hydrogen and carbonaceous wastes for the essential fuel cell function without the need for an external reformer or for pure hydrogen. Because of their high operating temperature, they are impervious to poisoning from carbon monoxide and other impurities—and they do not require expensive noble metals as catalysts.

In the United States, early commercial tests were conducted in the 1990s by the Santa Ana, California, utility with a 2-megawatt MCFC power plant built by FuelCell Energy's predecessor, the Energy Research Corporation, and of a plant built by MC Power at the Miramar Naval Station. Those early plants were fueled by natural gas.

Solid Oxide

Solid oxide fuel cells (SOFCs) were expected to come into their own as power generators for utilities shortly before 2010, but that has not happened commercially. The U.S. Fuel Cell Council's Web site lists more than 100 SOFC projects around the world, but they appear to be mostly research and development efforts of small stationary devices, and many of them are believed to have been decommissioned. Larger power plants are in the works by, notably, Siemens's Stationary Fuel Cells Division with support from the U.S. Department of Energy and the German Ministry of Economics and Labor; Siemens says on its Web site that it is now in the precommercial phase and expects to have its first commercial product in 2012. Other players in this category are Rolls-Royce, UTC Power, and FuelCell Energy. Developers of smaller stationary systems include Australia's Ceramic Fuel Cells, Acumentrics, Cummins Power Systems, Delphi Corporation, and Versa Power Systems. SOFCs They operate at the highest temperatures yet: 800°C to 1,000°C. Their efficiency is in the 55 to 65 percent range. The electrolyte typically consists of an alloy of rare earth materials, such as yttrium-stabilized zircon dioxide, a ceramic material. Small, even tiny SOFCs, such as the Lilliputian device mentioned earlier, are under development for a variety of applications for home use, various uses in remote locations (e.g., to power pumps for long-distance natural gas pipelines), and as onboard power generators for automobiles (taking the place of traditional generators and alternators). Like the molten carbonate type, they typically run on natural gas. Also under development are other advanced designs, such the direct methanol fuel cell (DMFC) and the reversible or regenerative fuel cell.

Probably the best-known early developers of DMFCs in the international fuel cell community were teams of scientists at the University of Southern California, including Nobel laureate George A. Olah, who is credited with inventing the DMFC at Jet Propulsion Laboratory and other organizations. The basic idea was to develop a type of fuel cell—an offshoot of PEM technology—that could be fed directly with methanol or a solution of methanol in water. This would be an advance over current types, in which hydrogen must be extracted from methanol in a miniature chemical plant (a "steam reformer" or a partial oxidation processor) before it can be fed to a fuel cell. That approach had technical drawbacks, such as the slight contamination of the produced hydrogen ("reformate" in engineering parlance) with traces of hydrocarbons. Developing a fuel cell robust enough to take methanol straight would "significantly lower system size, weight, complexity, and temperature than in existing fuel cell systems," said Gerald Halpert, lead scientist of the JPL effort, in an autumn 1996 paper.[23] In the late 1990s DaimlerChrysler, and possibly other carmakers. were working on DMFC propulsion but eventually gave that up in favor of hydrogen and PEM fuel cells Currently DMFCs are limited to small portable electronics applications, the most visible, energetic company being Germany's Smart Fuel Cells. In addition to Toshiba and its Dynario charger, DMFC developers include MTI Micro (in Albany, New York) and its Mobion system, Arkema, and others.

Reversible Fuel Cells

Systems that can be switched back and forth between producing electricity from the cold combustion of hydrogen and oxygen and splitting water again into hydrogen and oxygen by electrolysis were first investigated in the late 1980s and 1990s as laboratory exercises by the Hamilton Standard Division of United Technologies Corporation. Other researchers, notably Fred Mitlitsky at Lawrence Livermore National Laboratory and later at Proton Energy Systems, but also elsewhere, had been working on such concepts as well. (They emerged full-blown as the most bally-hooed launch in the fledgling fuel cell industry yet: the birth of the Bloom Box in early 2010. But I am getting ahead of myself.)

In the late 1990s, four of the Hamilton Standard experimenters set up a new company, Proton Energy Systems, which in summer 1999 announced it had successfully tested a "unitized regenerative fuel cell"—in effect, "a high-performance large-scale battery without many of the limitations of a conventional battery," the company said in an announcement. The concept was born in the early 1990s as a part of a NASA plan for a huge, unmanned, solar-powered flying wing. In order to stay aloft for months, the flying wing would have to store and recycle solar energy for night flight at stratospheric altitudes of 70,000 feet or more for military and civilian purposes such as long-term weather observation. The energy needed to crawl up to the fringes of space (a 1993 story in The Hydrogen Letter said that the plane would climb at a rate of only about 12 knots, requiring about 6 hours to reach 70,000 feet, and would then cruise at only 27 to 28 knots) would come from solar cells covering about 75 feet of the 100-foot wings' upper surface.[24] Operating as electrolyzers during the day, these onboard fuel cells would split water and would then use the hydrogen and the oxygen to produce electricity to turn the propellers at night.

Early on, the reversible fuel cell designers felt that it could have automotive applications. James McElroy, then program manager for electrochemical systems at Hamilton Standard, said in a 1993 interview that a reversible fuel cell "would be like any other chemical fuel cell, except that you wouldn't need a new infrastructure [for fuel]. You would plug it into the electric power grid as infrastructure." That has not happened yet, and regenerative fuel cells took a different path.

Then in 2010, a variant, the regenerative solid oxide fuel cell, emerged from the shadows of almost ten years of secret development by California's Bloom Energy Corporation as the modular Bloom's Energy Server, better known as the Bloom Box stationary power plant; Jim McElroy is one of Bloom's cofounders, and Fred Mitlitsky is one of its key engineers. The product of more than $400 million in venture capital funding, the Bloom Box was launched in a glittering February 2010 Sunnyvale, California, press conference at the headquarters of one of the first customers for the device, eBay, attended by Governor Arnold Schwarzenegger and former secretary of state and Bloom board member General Colin Powell. Other luminaries such as Senator Diane Feinstein, Silicon Valley entrepreneur Vinod Khosla, and New York mayor Michael Bloomberg paid tribute in video clips, and the festivities were preceded by Sunday night national prime-time TV interview with Bloom's principal cofounder, K. R. Sridhar. About fifty 100-kilowatt Bloom Boxes, fueled with natural gas at better than 50 percent efficiency, had been installed by industrial energy users such as Google, Walmart, Staples, Bank of America, and Coca Cola by September 2010. To keep the design simple for now and, at $700,000 to $800,000 a copy, affordable for industrial users, these units produce only electricity, but they can also produce hydrogen. CEO Sridhar said during the February press conference that he could envision solar panels on house roofs for producing both electricity for the home and hydrogen as car fuel. "This is extremely important to us," he said. The company plans to build smaller Bloom Box versions for home use, "the killer application in about a decade."[25]

Bloom Energy was not the only company working on regenerative solid oxide fuel cells. One competitor was Versa Power Systems of Littleton, Colorado, which in 2010 was working with Boeing and the Solid State Energy Conversion Alliance and with DoE funding on SOFCs. Another was UTC Power, which was skeptical about Bloom's progress and claims. Mike Brown, a UTC Power executive, was quoted in a February 24, 2010, *New York Times* story, cited as part of the Bloom Energy launch story in the March 2010 issue of *The Hydrogen & Fuel Cell Letter*, "We have been working with solid oxide for 30 years but we are still in the lab. . . . Nobody has been able to solve the reliability problems."[26]

Is there likely to be one dominant fuel cell technology? Probably not. Most likely, different types of fuel cells will be built for different purposes, with bottom-line economics as the determining factor. As pioneer Kordesch observed just before the turn of the century, "Economy will decide."

8

Clean Contrails: The Orient Express, Phantom Eye, and LAPCAT

It did not set any speed, distance, or altitude records, but it was an aviation landmark nonetheless: the world's first flight of a hydrogen fuel cell–powered airplane. The two-seater Austrian-built Dimona motor glider, converted to fuel cell power by Boeing's Madrid-based Research and Technology Europe group, climbed to a mere 1,000 meters (3,300 feet) and flew for about twenty minutes at a leisurely 100 kilometers per hour (62 miles per hour) in one of three test flights in February and March 2008. That was enough to set a new landmark. Boeing's researchers and engineers "have demonstrated the potential of integrating fuel cell technology into aerspace products and the promise of a brighter, greener future," said Boeing's senior vice president of engineering, operations and technology, John Tracy.[1] The output from the 25-kilowatt proton exchange membrane (PEM) fuel cell, made by Britain's Intelligent Energy, had to be augmented by a French-made SAFT lithium-ion battery to achieve the 75-kilowatt peak needed for takeoff, disconnected after the plane had reached cruising altitude. The plane, weighing some 100 kilograms more than the standard internal combustion–engine commercial version, carried only about 0.8 kilograms of hydrogen compressed to 350 bars (5,000 pounds per square inch) in a fueling system built by Air Liquide Spain.

A year and a half later, the German Aerospace Center, DLR, went Boeing one better with its Antares DLR-H2 single-seater fuel cell motor glider (figure 8.1). It achieved takeoff without a battery assist on fuel cell power alone during its eight-minute maiden flight from Hamburg's airport—a fraction of what its developers said the plane was capable of: carrying up to 4.9 kilograms of compressed gaseous hydrogen, cruising at 170 kilometers per hour (106 miles per hour), considerably faster than

Figure 8.1
A fuel cell motor glider, the Antares DLR-H2, capable of takeoff without a battery assist, was on display in fall 2008 suspended from the ceiling of Stuttgart airport's main concourse. It was designed by the German Aerospace Center DLR, derived from a high-performance motor glider built and sold by Lange Aviation, a German maker of motor gliders. An improved follow-up version with a slightly different layout was being readied for a possible attempt to cross the Atlantic on fuel cell power in 2012.

Boeing's plane, for up to five hours for a range of about 750 kilometers (468 miles). The power plant was a Danish 25-kilowatt maximum Serenergy PEM fuel cell that used a high-temperature membrane and catalysts made by Germany's BASF Fuel Cells; full power was needed for getting off the ground, but only 10 kilowatts were needed for level flight, according to DLR. At 20 meters, the Antares, built by Germany's Lange Aviation, had a slightly bigger wing span than the Boeing's 16.6 meters. The plane is essentially a flying test bed for DLR's project to develop a fuel cell auxiliary power system for large airplanes such as the Airbus A320. Beyond that, DLR intends to put the plane to use as a high-visibility PR instrument. Although this plan is not yet finalized, a high-level DLR executive let it be known during another flight demonstration that fall that the agency would attempt to cross the Atlantic—presumably from North America to Europe—with the fuel cell Antares, possibly in 2012.

Nor was it clear whether DLR would attempt a long nonstop flight—a fuel cell version of Charles Lindbergh's historic 1927 flight—or one in several stages

Completing the aerial fuel cell hat trick in mid-2010 in Italy was the European Union–sponsored ENFICA-FC (for "environmentally friendly inter-city aircraft powered by fuel cells") Rapid 200 fuel cell plane, a project coordinated by a researcher at Turin's Polytechnic Institute, Giulio Romeo. The two-seater, based on a Czech Republic–built Jihlavan Airplane Rapid 200, took off from an airstrip in Reggio Emilia, Italy, on its two-minute maiden flight in mid-May, followed quickly by other test flights, including one in which the designers claimed a world speed record of 135 kilometers per hour (84 miles per hour) in its class. Like Boeing's Dimona, it was powered by an Intelligent Energy fuel cell rated at 20 kilowatts, assisted for takeoff by a set of 20-kilowatt lithium-ion batteries. Like the Antares, the ENFICA-FC is part of a more ambitious project: a study designed to define new aircraft power systems for items such as auxiliary power units (APUs), primary electrical generation, emergency landing gear and other systems, including a more theoretical, feasibility study of an all-electric fuel cell–powered ten- to fifteen-seat commuter plane based on an Israeli Aircraft Industries twenty-seat commuter jet and a Czech design, the Evektor EV 55, as baseline configurations.[2]

These successful flights almost bore out the prediction ten years earlier by a NASA researcher, David Ercegovic, who led a seven-member team investigating zero carbon dioxide (CO_2) technologies in a three-year project at the Glenn Research Center in Cleveland, Ohio. He observed in an interview, "I am willing to say that a Cessna 172 class aircraft testbed [operating] on hydrogen and a fuel cell is not out of the question in the next five to ten years."[3] *Almost* is the operative word here: the Cessna is a four-seater and weighs a lot more than the ENFICA-FC, for example. Still, it is getting closer.

Hydrogen and fuel cells are becoming of interest in an off-the-beaten-track application: unmanned aerial vehicles (UAVs). *Fuel Cell Today's* 2010 Industry Review noted, for example, that UAVs were one of three niche transport markets that have become "breakout markets . . . which are experiencing solid market pull whether from government, military or direct consumer demand" (the other two are materials-handling vehicles and APUs.[4] Most of the interest comes from military establishments around the world but also from firefighters, telecommunication people, and others for missions such as surveillance of pipelines. UAVs are often preferred for missions that are too dull, dirty, or dangerous for manned aircraft.

At the large end of the UAV scale, Boeing made news in mid-2010 with another hydrogen airplane—the very large, bulbous Phantom Eye aerial surveillance plane. Boeing calls it an unmanned airborne system (HALE, for high altitude, long endurance) plane that in its initial two-thirds-sized demonstrator iteration can stay aloft for four days at altitudes of up to 65,000 feet and carry up to 450 pounds of surveillance and reconnaissance gear; larger HALEs are planned to keep flying for as long as ten days and carrying up to 2,000 pounds on shorter missions. A video of its unveiling in an auditorium in St. Louis, headquarters of Boeing's Phantom Works division, showed the plane's fat, short fuselage—most of it apparently housing a liquid hydrogen tank—towering about twice the height of speaker Darryl Davis, president of Phantom Works, above the floor. The plane's 150 foot wingspan is about three-quarters that of a Boeing 747 (195 feet). Unlike many other UAVs that run on fuel cells, the Phantom Eye is powered by two turbocharged 2.3-liter four-cylinder Ford Ranger truck engines, first modified by Ford ten years earlier to run on hydrogen, providing a cruise speed of about 172 miles per hour and a top speed of more than 230 miles per hour. Boeing has not said how much hydrogen is carried aloft or what the plane would weigh. But a May 2010 Boeing presentation found on the Internet, "Technology Developments to Support the BMDS" (Ballistic Missile Defense System), said the yet-to-come, bigger, full-sized version with a 250-foot wingspan would have a gross takeoff weight of 14,125 pounds. The plane was due to be shipped to NASA's Dryden Research Center at Edwards Air Force Base in California for ground and taxi tests, with the first flight scheduled for late summer 2011.

These kinds of ground and taxi tests were completed in early 2010 at Edwards for a similar hydrogen-fueled high-altitude plane in roughly the same league, the *Global Observer,* built by California-based AeroVironment. This unmanned plane, distinguished by a set of eight spindly propellers, is also designed to fly between 55,000 and 65,000 feet for up to a week at a time for surveillance and reconnaissance over any part of the world. The 2010 version was a one-third-scale demonstrator of the final version, powered by a fuel cell. Like its competitor Phantom Eye, the full-sized plane will be powered by a hydrogen internal combustion engine to generate electricity for the electrically driven propellers.

At the small end of the scale, AeroVironment has also developed a hand-launched 12.5-pound hybrid fuel cell unmanned plane, the Fuel Cell Puma, which in 2008 set a record by flying for more than nine hours. The PEM fuel cell, made by Protonex Technology Corporation in Southborough, Massachusetts, provides power during cruise flight and recharges the lithium ion battery that provides extra power for takeoff and for two cameras that send live streaming video images to the ground operator. Protonex also provided a 550-watt fuel cell for the small Ion Tiger unmanned aerial vehicle that in November 2009 set an unofficial endurance record by staying aloft for more than twenty-six hours. The 37-pound Ion Tiger, developed by the U.S. Naval Research Laboratory, carried a 5-pound payload, said the laboratory's release. Next steps include tripling the fuel cell's power to 1.5 kilowatts and extending flight times to three days.

Big Hopes—Dashed

As noteworthy as these projects are, they are a far cry from and a pale shadow of the high hopes and optimism for big hydrogen-fueled passenger planes the size of Boeing 747s, Lockheed TriStars, and Airbuses that many members of the fledgling international hydrogen community felt in the 1970s, 1980s, and 1990s. These hopes were highlighted by an event in April 1988 that did not get much coverage in the American media, but nevertheless was a landmark in the history of aviation and of hydrogen energy:

A modified Tupolev 154 airliner, renamed TU-155, took off from an airport near Moscow on a test flight with the turbofan engine on the right side operating on liquid hydrogen.[5] The English-language service of the Soviet news agency TASS reported that the engine could be operated on either liquid hydrogen or liquefied natural gas, and the newspaper *Izvestia* said the plane ran on liquid hydrogen during its maiden flight from an unidentified airport near Moscow.[6] Hydrogen, the story hinted, was expected to be the long-term fuel of choice for global aviation. That evening, the Soviet television news show *Vremya* showed the plane taking off and landing. In the United States, a brief segment was distributed by the CBS television network to local affiliates, but apparently few stations bothered to show it. The cable channel CNN showed footage of the flight three days later. The Washington publication *Defense Daily*, which did carry a brief story, reported that the flight had lasted twenty-one minutes. The *New York Times* noted the event about four weeks later.

To interested observers, it appeared that the Soviet Union was gearing up to develop hydrogen-fueled aviation and that it had a leg up over the West in exploring this new technology. Maybe it was only old-style communist propaganda, but both TASS and the party newspaper, *Izvestia,* implied that aircraft powered by liquid hydrogen and liquid natural gas would be the wave of the socialist future.

Even more startling to the international hydrogen community was that this was not a sudden public relations ploy. The Soviets had been at it for a long time. Alexei Tupolev, son of the world-famous aviation engineer who had created the series of Tupolev airliners and a brilliant aircraft designer in his own right, was quoted by *Izvestia* as saying that preparations for the flight had been going on for nine years and had been a priority for Soviet aviation research.

G. Daniel Brewer, a retired aeronautical engineer who had headed Lockheed's hydrogen program between 1972 and 1984, commented that the Tupolev team's achievements placed it ahead of U.S. efforts to develop the "national aerospace plane" (a hypersonic hydrogen-fueled experimental craft, announced by President Reagan two years earlier, that was to fly from a normal runway to beyond the atmosphere and back). Brewer told the *Defense Daily* reporter that the Russians might be "five years ahead of us in a technology that we need to develop." "Since the Russians have flown [their] plane," he said in another interview, "it possibly means that they have done all the [cryogenic] valves, heat exchangers, pumps." On the other hand, "if the plane used a pressure-fed liquid hydrogen system as we did in the late 1950s, it wouldn't be all that great," he said. (He was referring to the fact that in 1957, an American B-57 twin-jet bomber had flown many experimental missions, of about twenty minutes each, with one engine operating on liquid hydrogen.) But "if they have that system to take off, fly and land (on pump-fed liquid hydrogen) that would really be a step forward," Brewer added. "It would take us at least five years to develop it and put it in operation to achieve the same point."[7]

At the World Hydrogen Energy Conference held in September 1988 in Moscow, Tupolev told an overflow audience that hydrogen had been used exclusively to power the right engine of the TU-155 throughout the entire flight cycle—takeoff, cruising, and landing. He said that the liquid hydrogen had been carried in a removable stainless-steel tank that had taken up most of the plane's rear compartment and that there had been enough for a flight of up to ninety minutes. Potential ignition sources—electrical wires, hydraulic lines, conventional fuel lines—had been removed and relocated elsewhere in the fuselage. Pressurized air had been blown into those areas and into the space between two walls that separated the rear compartment from the rest of the plane to prevent the buildup of potentially dangerous air-hydrogen mixtures.

The Soviet feat caused muted anguish in Washington among the few lawmakers who supported the development of hydrogen energy technology. Tupolev's triumph was "yet another case in which the US had stood idly by in a critical area of research and allowed other nations to move far ahead," the late Representative George Brown Jr. (D, California), a member of the House Science, Space and Technology Committee, said shortly after the flight. "The Soviets are not alone in the development of liquid hydrogen as a fuel for their transportation. Now, with the Soviet cryogenic program, we face a real national security threat. Judging by the lack of support for hydrogen R&D in this country, it appears that the United States will be forfeiting to other more farsighted nations."[8] Two weeks later, Senator Spark Matsunaga, Brown's counterpart in the Senate and a frequent cosponsor of hydrogen legislation, called the Soviet flight "a milestone." In an April 27 speech on the Senate floor, Matsunaga urged Congress and the Reagan administration, both monumentally uninterested in cleaning up the environment or in alternative renewable energy, to support hydrogen legislation that he and Brown had introduced earlier and had been languishing in committee. "It is not too late to move on this bill and pass it before the end of this Congressional session," Matsunaga said. "As the news from Moscow indicates, my bill will come none too soon for international competition in aviation and space."

Thirty years earlier, *Sputnik* had "shocked and galvanized our nation," Matsunaga told his colleagues. "Eleven days ago, the Soviets announced another scientific 'first' which, curiously enough, has gone virtually unnoticed to date in this country." The flight of a jet plane "powered by liquid hydrogen in an engine modified to accept cryogenic fuel . . . spells the advent of cryogenic aviation and represents a milestone in the march toward a hydrogen economy." Given the importance that the current administration had placed on the development of a "trans-atmospheric aircraft" and "the recognition that hydrogen would be the fuel of choice for such a craft," Matsunaga said pointedly, "it is a mystery to me why the administration has neither welcomed nor supported my legislation. Perhaps now that the Soviets appear to have again stolen a technological march on us in this regard, administration officials might be moved to reconsider their position on my bill."

Presumably unbeknown to Senator Matsunaga when he made his speech, an American aviation pioneer and veteran pilot was quietly getting ready to claim the hydrogen aviation spotlight (or at least a small part of it) for the United States. Working almost alone, William Conrad, an octogenarian retired air transport ratings examiner who had started flying in 1929 and had been Pan American's first director of flight training, was at work in a hangar at Fort Lauderdale's Executive Airport, preparing a second-hand four-seat Grumman Cheetah to take off, fly, and land while running exclusively on hydrogen.

On June 19, 1988, Conrad, a recipient of the Wright Brothers Memorial Award and an inductee in the Aviation Pioneers Hall of Fame before this flight, got up at 5:00 A.M., drove some 60 miles north to West Palm Beach, filled his custom-made 40-gallon portable cryogenic tank with liquid hydrogen purchased at $2.80 a gallon from the commercial supplier Tri-Gas, and drove back to Executive Airport. There, he and his mechanic used a mobile crane to lower the tank into the cockpit behind the pilot's seat and connected the fuel lines to the plane's custom-made hydrogen fuel system. Inspectors from the National Aviation Association NAA) and the World Air Sports Federation (Fédération Aéronautique Internationale—FAI) were on hand to witness the event; they certified that the gas tanks in the wings had been sealed off. Conrad revved the engine and checked it out "to make sure that everything was all right," strapped himself in his seat, slid the cockpit canopy shut, and was ready to go. He then received clearance from the tower for what was supposed to be only a high-speed taxi run. (As he related later, fear and ignorance of hydrogen was such that he never managed to get permission for a full test flight.) At about 1:30 P.M., Conrad pushed the throttle lever full forward. The Lycoming E2G engine, rated at 150 horsepower on gasoline, produced about 10 percent more power on hydrogen.[9] About 20 seconds later and some 600 to 700 feet down runway 8, the Cheetah became airborne. It rose about 300 feet, then touched down safely. The flight had lasted 39 seconds, never leaving the airspace above the 8,000 foot runway. "The damn thing just got into the air with me," the tall, lanky Conrad deadpanned afterward. Although the hop was technically only a high-speed taxi test, it was enough for NAA and FIA inspectors to certify a world first for Conrad.

Conrad, who had spent about $100,000 on the project (including $26,000 of his own money for the plane itself), said that he "just wanted to establish the fact that the United States has the first plane to run solely on hydrogen." The purpose, he added, was "to get a little publicity in order to make people realize that hydrogen is available and that it is a nonpolluting form of energy."

President Reagan sent Conrad a congratulatory letter later that year. "Your remarkable flight last June was a milestone in aviation history, and your 14 years of preparation for it were welcome proof of the vitality of our search for new energy sources to improve transportation efficiency and benefit the environment," he wrote.

Conrad died about a year after the flight, on June 27, 1989, after a long bout with cancer. His passing was noted by Senator Matsunaga "with a heavy heart" in a September 19 speech in which he urged his fellow senators to support research on the use of hydrogen as an aviation fuel. "We must not lose the initiative in hydrogen-powered flight that Bill Conrad set for our country," he said. Tupolev's and Conrad's accomplishments marked significant advances over military efforts in this area two decades earlier. In 1957, an American B-57 twin-engine jet bomber cruised at about 50,000 feet over Lake Erie at about Mach 0.75. It looked just like any other B-57, except perhaps for the wingtip tanks: the one on the left was slightly thinner and less smoothly rounded than the one on the right, and it looked a bit makeshift. Another difference, almost unnoticeable, was a small boxlike structure mounted outboard of the left engine underneath the wing. It was an air-hydrogen heat exchanger.

Based at the NACA Lewis Research Center, near Cleveland, this B-57 was the first airplane ever to fly partially powered by liquid hydrogen. After taking off on kerosene, the pilot could switch one engine to draw pressurized liquid hydrogen from the left wingtip tank. Pressure provided by helium carried in the right wingtip tank would push the liquid hydrogen to the heat exchanger, where heat from the inrushing air would turn it into a gas. The hydrogen gas could then be burned quite normally in the left engine. The plane flew for about two years, using hydrogen in flights as long as 17 minutes, always at high altitudes and at cruising speed, with no fuss and no problems.

Also in the 1950s, studies conducted by the Lockheed Corporation in collaboration with the engine maker Pratt & Whitney and a division of AiResearch led to a remarkable episode in the annals of hydrogen technology: the secret project to develop a liquid-hydrogen-fueled supersonic spy plane, the CL-400 (code-named Suntan). The story of this project was closely guarded until 1973, when former project engineer Ben Rich described the never-built plane (it never reached the prototype stage) to an audience of aviation and hydrogen experts at a hydrogen aircraft symposium at NASA's Langley Research Center.[10] That was the first time anything about this project, which eventually led to the famed SR-71 was described to outsiders.

As Rich told it, the CL-400 was intended to fly at close to 100,000 feet, to cruise at Mach 2.5, and to have a range of 1,100 miles. The first of two prototypes was to be delivered only a year and a half after the go-ahead. The CL-400 was supposed to carry a crew of two and a payload (mostly aerial cameras and other reconnaissance equipment) of only 1,500 pounds. More than 164 feet in length, it was to have a wingspan of about 84 feet. The fuselage was to be 10 feet in diameter. The two engines, each delivering 9,500 pounds of thrust, were to be mounted at the wingtips, where some other planes carried auxiliary fuel tanks. Small outrigger landing gears at the wingtips would stabilize the plane on the ground. "The low density of the cryogenic fuel necessitated a large fuel volume," Rich recalled—21,404 pounds of liquid hydrogen. At takeoff, the plane would have weighed 69,955 pounds.

"There was very little knowledge of liquid hydrogen or cryogenic handling, other than that related to the hydrogen bomb experience," Rich said at the Langley symposium. "Consequently, it was necessary for us . . . to set up a test facility to learn how to handle liquid hydrogen in a fashion no different from hydrocarbon fuel. It was our feeling that if you could not handle liquid hydrogen like gasoline, you did not have a practical vehicle." Under the overall direction of Clarence Johnson, the director of Lockheed's advanced projects organization, the Skunk Works, Rich and his crew set up small-scale hydrogen tanks and fuel supply systems to get the hang of handling the stuff. Much effort was spent on discovering how to run liquid hydrogen at −425°F through wings heated to several hundred degrees by air friction at supersonic speeds.

A great deal of work also went into studying the much-feared but largely unknown hazards of hydrogen, including a large number of deliberate attempts to explode it. "The deflagrations were generally mild due to the high hydrogen flame velocity," Rich said. "The fireball was much less than a comparable kerosene fire. . . . Only 2 of 61 liquid hydrogen tests produced bona fide explosions. In both cases oxygen was deliberately mixed with liquid hydrogen"—just about the most dangerous situation imaginable and not very likely in normal operations. Said Rich: "We showed that a hydrogen aircraft was feasible. Liquid hydrogen could be handled, with the proper procedures and care, as easily as hydrocarbon fuel."

The CL-400 got as far as wind-tunnel testing, component development, and procurement of basic materials, but in 1957, the program was canceled. A technical reason for the cancellation was that once the overall design was fixed, there was not much of an opportunity to improve the plane's range, mainly because of the peculiar characteristics of liquid hydrogen. "The airplane was too short-legged and had not more than 5 percent potential range performance stretch," according to Rich—the result of the design's low supersonic lift-to-drag ratio. "It grew into a big dog without enough range," Rich noted in a 1994 *Popular Science* interview.[10] Another reason for the cancellation had to do with the logistics of transporting liquid hydrogen (LH_2) to air bases on the periphery of the Soviet Union and China. "How do you justify hauling enough LH_2 around the world to exploit a short-range airplane?" Rich asked. Other studies showed that switching to hydrocarbon fuels could double the plane's range; ultimately the decision to do so led to the SR-71 supersonic reconnaisance plane.

According to John Pike, the aviation expert of the Federation of American Scientists, although Suntan was a failure, "the work done on its hydrogen propulsion system laid the groundwork for subsequent application of this technology for space rocket propulsion." The Suntan program led to the Centaur (the first space rocket fueled with liquid hydrogen), the *Apollo* lunar program, and the space shuttle.

Pound for pound, liquid hydrogen stores about 2.8 times as much energy as jet-grade kerosene. At the same time, however, 1 pound of liquid hydrogen takes up three to four times as much volume as a pound of jet-grade kerosene. A pound of liquid hydrogen has a heating value of 51,500 Btu. Jet-A, the most common jet fuel, has only 18,600 Btu per pound. But a cubic foot of liquid hydrogen has only 227,700 Btu, whereas the same volume of Jet-A has 906,000 Btu. Thus, to obtain the same heating value contained in a cubic foot of Jet-A requires 3.97 cubic feet of liquid hydrogen. For this reason, hydrogen-fueled aircraft will have to have "fat" fuselages; however, they will be lighter than kerosene-fueled planes.

Quite apart from the environmental advantages of burning hydrogen in jet engines (which are basically the same as those of burning hydrogen in car engines: no smoke, no unburned hydrocarbons, no CO_2 or CO, low nitrous oxide emissions), there may be economic advantages.

At the same 1973 NASA hydrogen symposium at which Ben Rich described the Suntan spy plane, two NASA scientists, Cornelius Driver and Tom Bonner Jr., presented the first rough estimates of what hydrogen use might mean in terms of a practical design goal: a jet freighter capable of carrying 265,000 pounds over 5,000 nautical miles. Their study did not assume any design advantages that might come about as the result of using hydrogen; they merely substituted liquid hydrogen numbers for kerosene data in standard design computations. Nonetheless, the numbers were startling. Using the general formula that provides a measure of an airplane's overall flight efficiency,

$$\frac{M(L/D)}{SFC}$$

where M is Mach number, L is lift, D is drag, and *SFC* is specific fuel consumption, Driver and Bonner said that using hydrogen in planes such as the 707, the DC-8, the 747, the DC-10, and the L-1011 would roughly double fuel efficiency "if no serious problem is encountered in providing the necessary hydrogen volume."

In 1976 Lockheed's Brewer came up with more refined data on both supersonic and subsonic hydrogen planes. Of twenty-four designs that had been studied, Brewer discussed a long-range subsonic passenger plane (400 passengers, range of 5,500 nautical miles) as a representative example.[11] The data were encouraging. Gross weight was 177 tons for the hydrogen plane versus 237 tons for the Jet-A version. Fuel weight for the hydrogen plane was one-third lower, and the engines could be smaller and lighter since they would have to deliver 11 percent less thrust. Wing area was reduced by about one-fourth because of the lighter loads, though the hydrogen plane was about 10 percent longer. The plane would carry liquid-hydrogen tanks inside the fuselage fore and aft, with no direct access from the passenger compartment to the cockpit.[12] Because the forward hydrogen tank would be in the way, a potential hijacker would be unable to reach the cockpit in flight. The hydrogen-fueled subsonic transport would be considerably quieter during takeoff and cruising but not during landing, the analysis predicted. For flyover noise, the hydrogen plane would register 104.9 effective perceived noise level in decibels (EPNdB); the Jet-A plane would register 107 EPNdB. On a landing approach, the liquid hydrogen plane would be somewhat noisier: with most of its fuel gone, a hydrogen plane would weigh about the same as the similarly empty Jet-A plane; however, because it would have smaller engines and smaller wings, the engines would have to run at a higher power setting to maintain the same glide angle, which would make more noise. The liquid hydrogen plane would affect a smaller area with high noise—the so-called 90 EPNdB contour, the area around the runway in which these noise levels are registered during takeoff and landing—than an equivalent kerosene-fueled plane. The hydrogen version would produce no carbon oxides, unburned hydrocarbons, or smoke, but it would emit water vapor and some nitrogen oxides. In cruise conditions, Brewer estimated, hydrogen engines would produce about twice the amount of water vapor than the kerosene plane—82.4 pounds per nautical mile versus 41.9 pounds. Would these water vapor emissions affect the weather? In the absence of experimental data, it was hard to say, but Brewer felt that the meteorological impact would be minimal or nonexistent. If the water vapor coming out of the plane's four jet engines were visualized as a thin film of water the width of the engines' exhaust nozzles, the thickness of this water film would be only 0.00008 inch, and 82.4 pounds of water spread over a nautical mile does not seem very worrisome.

What about nitrogen oxides, the by-products of any burning in the atmosphere? With hydrogen, chances are that the problem can be diminished, if not licked. At the 1973 NASA symposium, two scientists from New York University, Antonio Ferri and Anthony Agnone, said that existing types of turbojet engines burning hydrogen would actually produce more nitrogen oxides. But, they added, the situation would change completely if different combustion schemes (such as vaporizing the hydrogen, premixing it with air, and injecting it into the combustor before burning) were introduced.[13] Ferri and Agnone concluded that in future subsonic planes, the formation of nitric oxides could be much reduced by combustor designs of this type.

If hydrogen makes subsonic planes look good, its use in a future supersonic transport (SST) glows with economic promise and environmental health. A proposed American SST, which Boeing had planned to build after winning a design competition against Lockheed, was killed by environmentalists in the U.S. Senate in 1971. In retrospect, this was probably a good thing. Aside from pollution concerns, the plane's marginal economics probably would have sounded its death knell rather quickly after the Arab oil embargo and the energy crisis of 1973. The French-British Concorde never became a big commercial success—its last flight was in 2003—and the TU-144 (a Soviet SST that looked much like the Concorde) was mothballed within a few years because of technical problems.

When it comes to emissions, hydrogen-fueled SSTs would have the same advantages as subsonic hydrogen-burning airplanes. They would produce no carbon dioxide, carbon oxide, or unburned hydrocarbons at all, and the potential exists to lick the nitrogen oxide problem with advanced combustion technologies. Sonic booms could be reduced by design and by the fact that high-speed supersonic transports would likely fly much higher than today's jets (a Mach 6 plane might cruise at 100,000 feet) with less of the sonic boom shock reaching the ground.

Quick-turnaround, high-speed, or very large aircraft are likely to be needed to minimize congestion in the air and around airports. The International Air Transport Association projected in its 1997 long-term forecast that the volume of international air passengers would increase from 409 million in 1996 to 948 million by 2011; air freight and charter flights are also expected to grow substantially.

Making the case for a hydrogen SST, Brewer wrote in the mid-1970s, the same characteristics that made liquid hydrogen superior to kerosene for subsonic planes apply to an SST—only more so. Using the Mach 2.7, 234-passenger, 4,200-nautical-mile basic configuration with which Lockheed lost to Boeing in the 1971 competition, Brewer found the hydrogen-fueled version was longer but also almost one-third narrower. (Narrowness would facilitate taxiing and parking in congested airports.) The weight savings for the liquid hydrogen version would be stupendous. At 167 tons gross weight, the plane would weigh only half as much as an equivalent Jet-A plane. It would need only one-fourth the fuel weight of the Jet-A plane to fly the same distance, and the engines would have to provide little more than half as much thrust. For identical payload, range, and speed, the liquid hydrogen plane would have to carry 37 tons of fuel, the Jet-A plane 148 tons. Thus, the liquid hydrogen SST would also be more energy efficient. According to Brewer, it would require 4,272 Btu per seat per mile. The Jet-A design would require 6,102 Btu— 43 percent more.

The environmental aspects that made the hydrogen-fueled subsonic transport plane attractive also applied to the liquid hydrogen–fueled supersonic plane. Both sideline and flyover noise were expected to be lower because the plane would be considerably lighter. Sonic boom pressures would likely be lower because of the liquid hydrogen plane's reduced weight and smaller wing area. Further pressure reductions were believed to be possible with advanced designs.

A unique characteristic of liquid hydrogen is its ability to absorb large amounts of heat. For a plane flying at multiples of the speed of sound, liquid hydrogen's vast cooling capability becomes an important additional asset. For a Mach 6 plane zooming along at 90,000 feet, for instance, the friction of the hypersonic airstream builds up temperatures of up to 2,500°F at the nose and at the leading edges of the fuselage, the wings, and the control surfaces. Ultra-cold liquid hydrogen would be funneled through carefully spaced cooling tubes through all the critical areas where heat buildup occurs; it would also cool the flight deck and the passenger compartment. A secondary liquid might be circulated through the airframe and the wings to carry heat to a heat exchanger, where it would be transferred to the liquid hydrogen to impart an extra measure of energy before combustion in the engines.

In sum, then, the verdict regarding the use of liquid hydrogen for long-range subsonic and supersonic commercial aircraft, is, in Brewer's words, "an enthusiastic yes."

Almost immediately that view ran into criticism, some of it quite severe. Some critics have argued that in terms of total cost—including the expenses of production, transportation, and storage—liquid methane may be less expensive than, and therefore economically preferable to, liquid hydrogen. (Similar arguments are made today in favor of biofuels or synthetic liquid fuels.)

For reasons of safety and availability as well as for environmental reasons, Brewer dismissed liquid methane as an inferior choice. He argued that since civil commercial aviation is a truly international business, with airplanes refueling in New York, Rome, Tokyo, Jeddah, Warsaw, and Karachi, a future synthetic aviation fuel must be equally international—locally producible, with uniform characteristics everywhere. This is no problem as long as petroleum is cheap, and at the time kerosene and other hydrocarbon fuels were available at low cost all over the world. But aircraft makers, who have to think decades ahead when planning the useful life of airplanes, must face the fact that beginning sometime in the period 2010 to 2020, petroleum may cease to be cheap. Brewer contended that it was going to be necessary to develop a new generation of synthetic aviation fuels that would be available everywhere around the world. For Brewer, the answer was clearly hydrogen, as he observed in 1976:[14]

If international air travel is to continue to flourish and expand as projected in the face of definite prospects that some countries may be unable to obtain adequate supplies of petroleum at all times, it becomes mandatory either that all nations agree to share their petroleum fuel supplies or that they adopt an alternate fuel that can be commonly produced without hazard of control by a cartel.

Hydrogen offers many potential advantages for this application including the facts that (1) it can be manufactured from coal and water, or from water directly, using any of several processes and a wide variety of possible energy sources, and therefore can be considered to be free of the dangers of cartelization and (2) used as a fuel for aircraft it has been shown to provide significant improvements in vehicle weight, performance, and cost, and to result in reduced pollution of the environment.

As it turned out, liquid hydrogen aircraft propulsion technology went nowhere fast. Lockheed's interest in hydrogen as a fuel for jet aircraft waned with the retirement of Brewer in 1985 and of his boss, Willis Hawkins, in 1993. Lockheed could have beaten the 1988 Tupolev hydrogen plane by about ten years with government and industry support. As early as 1978, Lockheed had begun cautiously floating the idea of a small experimental cargo airline linking the United States, Europe, and the Middle East and employing four wide-body Tri-Star jets converted to liquid hydrogen. Lockheed felt that this liquid-hydrogen experimental airline project (LEAP) would convincingly demonstrate the technical, environmental, and potential economic advantages of liquid hydrogen as a jet fuel. As was outlined at the World Hydrogen Energy Conference in 1978 in Zurich,[15] Lockheed wanted to deploy the planes on a circuit linking Pittsburgh, Frankfurt, Riyadh, and Birmingham, England. A September 1979 International Symposium on Hydrogen in Air Transportation was organized to come up with practical ideas for LEAP. A comprehensive technology-development program was developed, but LEAP failed to win wide acceptance; indeed, *Business Week* reported that support for it ranged "from negative to lukewarm."[16] An ad hoc executive group with representatives from the United States, Canada, France, Germany, Belgium, Japan, and Saudi Arabia continued to meet and plan for more than a year afterward, but the project faded away.

In Europe, the story was more encouraging, but in the end, the result was the same. In March 1989, a little less than a year after the Tupolev flight, the German newspaper *Welt am Sonntag* reported that the German aerospace manufacturer Messerschmitt-Boelkow-Blohm (MBB) had decided to convert an Airbus to run on liquid hydrogen, and that the plane was to be operational within seven years. The paper even published an illustration. The plane would have a range of about 1,000 nautical miles. In June 1989, at the Paris Air Show, MBB confirmed that it was "seriously" considering conversion of an Airbus to liquid hydrogen, that a working group consisting of a dozen companies was gearing up for a feasibility study, and that there were hopes for a first test flight in the mid-1990s. In May 1990, at the Hannover Air Show, the Soviet Union and West Germany, two countries in the midst of dramatic political transitions, signed a preliminary agreement to jointly develop a "Cryoplane" for civilian purposes, The original Russian liquid-hydrogen plane, the TU-155, was on display—apparently the first time that the plane was shown in the West. It had logged fifteen hours flying on liquid hydrogen and forty hours on liquid methane, according to a May 28 story in *Aviation Week and Space Technology*. "There are no major technical obstacles for the use of liquid hydrogen or liquid methane for transports," Vladimir Andreev, chief designer for the Tupolev Design Bureau, told the magazine. "Our first tests were with liquid hydrogen, and we now are using natural gas—but if we wanted to switch back to liquid hydrogen, we can do this without a problem."

The basic idea was to convert either a Tupolev jet or an Airbus to run entirely on liquid hydrogen; the Germans pushed for the Airbus because it would be easier to sell in Western markets. The most visible design change would have been the addition of a hump to house the liquid-hydrogen tanks. It would add drag and degrade performance, but Airbus engineers believed it would be preferable to either redesigning the wings (the traditional place for fuel tanks) or carrying the fuel in two tanks, one in the front and one in the rear of the fuselage, as American designers had proposed earlier. Airbus engineers calculated that the use of super-light liquid hydrogen as a fuel could add valuable revenue-earning capacity for cargo or passengers, or both.

In 1990 the partners in the Cryoplane project hoped that a liquid hydrogen–fueled Airbus or Tupolev TU-154 could be flying in about five years. That was not to be. Two years later, Airbus concluded there was no justification for building what would be a very expensive hydrogen-powered demonstrator of that size. Instead, the partners decided to focus on the development of combustion chambers, pumps, valves, tanks, and other components.

In 1994 the project took a new tack with the news that Daimler-Benz Aerospace Airbus (DASA), the new corporate entity that had absorbed MBB and its Deutsche Airbus division, was now considering conversion of a smaller commuter plane to liquid hydrogen rather than an Airbus. That plane would be a DO-328, a twin-engine turboprop plane made by Dornier, a venerable company that Daimler-Benz had acquired earlier. The main reason for the switch was that converting a smaller plane would cost less than converting a big jet transport. Work on the plane, expected to cost about 60 million deutsche marks ($38.7 million at the time), was scheduled to start in January 1997 despite some uncertainties about financing and lack of a final signoff on the project on part of top DASA management.

The project was made potentially more attractive commercially by a change of engines. In summer 1996, 80 percent of Dornier's shares were acquired from Daimler-Benz Aerospace by the Fairchild Corporation. To make the plane more attractive to commuter airlines, Fairchild opted to switch to jets from the original turboprops. DASA, which kept one plane for the liquid-hydrogen project, decided that it made sense to go with jets for the hydrogen version as well but to leave other aspects of the plane—for example, wings and control surfaces—essentially intact. Two engine makers, Pratt & Whitney Canada and Allied Signal, had initially expressed interest in supplying a power plant.

In the end, though, all this work and plans came to naught. In early 1999, management decided to put the project on hold indefinitely because of financing problems: a shortfall of 10 to 15 million deutsche marks ($5.3–$8.8 million).

DASA, Tupolev, and their allies were almost the only aircraft producers taking any interest in hydrogen as an aviation fuel, but others showed no visible interest at that time—least of all Boeing, the giant among the world's aircraft manufacturers. A Boeing contribution to the 1993 *Transportation and Global Climate Change*, published by the American Council for an Energy-Efficient Economy, reiterated the company's long-held position that hydrogen was something to be considered only for the distant future. One of the chapters in it, "Characteristics of Future Aviation Fuels," authored by O. J. Hadaller and A. M. Momenthy, acknowledged an increasing number of environmental concerns, such as global warming, driving the search for a replacement of petroleum-based fuels. However, studies indicate that the currently used aviation fuel is as likely to satisfy these concerns as the few alternative fuels that are suitable for use in aircraft. Hydrogen . . . will become economically acceptable only after the world has exhausted its fossil fuel resources or a low-cost, abundantly available source of electric power, such as nuclear fusion, is developed. . . . Improving efficiency will be the principal way to lessen the impact of aircraft on the environment until a technically and economically practical non–fossil based fuel is discovered.

NASA still showed some interest. In some officials' eyes, hydrogen stood out as a prime candidate. One of them was Richard Niedzwiecki of NASA's Lewis Research Center who in the mid-1990s headed a section that worked principally with gas turbines. His scientists investigated combustion design and fuels and their emissions as part of atmospheric research. Briefing the Department of Energy's Hydrogen Technical Advisory Panel in November 1996, Niedzwiecki said that European countries were in the forefront of assessing the impact of aviation on the atmosphere. Looking at all possible fuel alternatives that do not produce CO_2, Niedzwiecki said, "The only thing I can come up with that is viable as a fuel is hydrogen. . . . Hydrogen is an excellent fuel for many applications, including aircraft." On the question of whether to tap into fossil fuels in the transition or whether to go directly to splitting water as a hydrogen source, Niedzwiecki left no doubt where he stood: "Environmentally acceptable manufacture . . . is the most important thing of all. I've heard here today that we could make hydrogen out of different [hydrocarbon] fuels, and then take the hydrogen and use it. . . . I might suggest an atmospheric scientist would wince at that. The general conclusion is, CO_2 is CO_2. . . . It doesn't matter if you produce it on the ground in Timbuktu or you produce it at 90,000 feet right over the White House." Either way, CO_2 would be injected into the atmosphere. Niedzwiecki continued: "I would strongly suggest that programs in the area of making hydrogen from non-hydrocarbon fuels would certainly be a priority of mine." Combusting hydrogen in a jet engine would produce almost three times the amount of water in the atmosphere than burning kerosene-type aviation fuel to produce comparable thrust, said Cecil Marek, a technical advisor to a NASA team at the Glenn Research Center that was looking at hydrogen fuel for airplanes. But kerosene produces 25 percent more of other emissions, and "water washes out of the atmosphere a lot faster than carbon dioxide which stays up for years." Still, injecting more water into the atmosphere creates more uncertainties about the formation of high-altitude cirrus clouds and their effects, all of which would need more study, according to Niedzwiecki. But "any atmospheric scientist you talk to would say that's a thoroughly acceptable trade at the present time."

Niedzwiecki acknowledged that the issue of high hydrogen costs versus current low jet fuel costs was going to be fiendishly difficult to deal with. Many people in the industry would resist switching to a revolutionary new fuel such as hydrogen because of fear of escalating costs and because they want to stay in business, he said. But "if it's an environmental issue, and if it leads towards the slowing down or the elimination of growth in the aircraft industry, that cost will not be a discriminator [for] people that work in those fields, and I think [it] probably would be gladly accepted." That seems not to have happened.

Lockheed veteran Brewer who also attended that meeting and who in twelve years as Lockheed's hydrogen manager produced "several million dollars' worth of studies for subsonic, supersonic and hypersonic" airplanes, said that a technology development program should be launched that would draw on the planning work done at Lockheed and elsewhere. But the prospects did not look good, he acknowledged: "In spite of these advantages, the overall support for use of hydrogen by our government— that is, NASA, FAA and all the other government agencies that are involved in this choice—is nil."

New Hope? The Green Freighter Study

Given the new century's accelerating climate worries, there could be a revival of interest in commercial hydrogen airplanes. In 2010, a four-year study that looked at the prospects for hydrogen-powered transport planes was presented at an international aeronautical sciences congress in Nice, France. Although there are no plans to build such planes, one study, "Hydrogen Powered Freighter Aircraft—The Final Results of the Green Freighter Project," examined a hydrogen version of the twin-turboprop ATR 72 regional freighter built by a French manufacturer, Avions de Transport Regional (figure 8.2). The other is a futuristic blended-wing body airplane that exists only as a design study, predicated

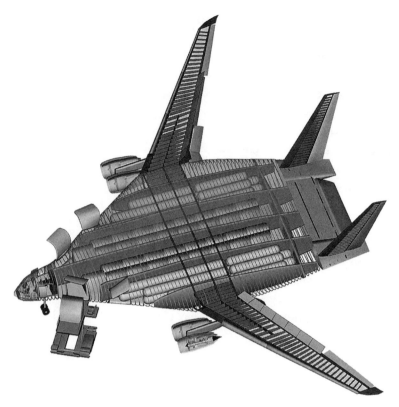

Figure 8.2
A rendering of a conceptual four-engine large hydrogen-powered "green freighter"
long-distance cargo plane, developed by researchers at Hamburg University of
Applied Sciences; Braunschweig Technical University; Airbus; and Bishop GmbH
Aeronautical Engineers, Hamburg. The study was presented at an Aeronautical
Congress in Nice, France, in September 2010.

on roughly the same size and payload as the twin-jet Boeing B777F
freighter. Its authors and principal investigators, Kolja Seeckt and Dieter
Scholz of Hamburg University of Applied Sciences, and Wolfgang Heinze
of Braunschweig Technical University, concluded that hydrogen planes
would offer technical advantages such as lower weight and shorter takeoff
runs but would not offer any better economics. Other partners in the
study were Airbus and Bishop GmbH Aeronautical Engineers, Hamburg
(details are on the Green Freighter Web site, http://GF.ProfScholz.de).

One of the reasons that the investigators decided to look at cargo
planes rather than passenger planes was that air cargo traffic is forecast

by both Airbus and Boeing to grow somewhat faster in the next twenty years than airline traffic: 5.2 percent annually and 5.4 percent, respectively, by Airbus and by Boeing for cargo, and 4.7 percent and 4.9 percent for airline traffic. Also, they believed tackling freight traffic first would be easier and more practical because air cargo transport is limited to a relatively small number of airports, making the needed infrastructure changes simpler.

Analysis of the smaller regional ATR hydrogen freighter found that it would have about a 5 percent smaller takeoff weight of 22 tons, including an 8-ton payload, and consume about 10 percent less energy than the operational kerosene-fueled version, despite the fact that when empty, the plane would weigh about 7 percent more. Calculations for a four-engine blended-wing-body long-range freighter of roughly the same payload—108 tons—as the twin-jet Boeing B777F indicated the liquid hydrogen plane would have a takeoff weight of 310 tons, or about 7 percent less than a hypothetical kerosene version, and the plane would burn 33 tons of hydrogen for the design range of 4,779 nautical miles (5,500 miles). Aside from the main benefit of producing very little or no carbon dioxide emissions and the presumably much reduced dependence on fuel from oil-exporting nations, the study again confirmed liquid hydrogen's big technical advantage of much lower maximum takeoff weight, crucial in aviation, compared to kerosene. But there are also important disadvantages: hydrogen is expensive to produce and is about four times as voluminous as kerosene for the same amount of energy—all of which means a hydrogen fuel system would be very heavy because of the large tanks and thick thermal insulation.

Concluding, the researchers said the lower takeoff weights would translate into takeoff runs between 14 and 28 percent shorter than with the kerosene versions. But these advantages do not translate into economic benefits, assuming the cost of both fuels in the future will be equivalent in terms of energy content: the hydrogen plane is expected to have 15 percent higher direct operating costs because of their higher empty weights.

Another sign that interest in hydrogen as aviation fuel may be coming back is a new patent granted to Boeing in early 2011 for a ring-shaped liquid hydrogen tank that would fit into a futuristic blended wing-body airplane design, a concept the company has been exploring for a long time. The patent for a "hydrogen fueled blended wing body ring tank" (patent no. 7,871,042), applied for four years earlier, may have been one by-product and outcome of awards NASA made in late 2010 to three teams—Lockheed Martin, Northrop Grumman, and Boeing—to study advanced-concept designs for aircraft that could start flying in 2025. The tank would encircle the roomy, almost theater-like interior cargo or passenger space of such a plane, and it would weigh less than conventional tanks for a given amount of liquid hydrogen because of the absence of the heavy end domes found on cylindrical pressure tanks. The Boeing tank would avoid changing the plane's aerodynamic shape, a problem very much in evidence with the hump-backed Airbus Cryoplane, and it would not take away usable passenger or cargo area in the plane, as in the Russian Tupolev 155. The continuous ring shape "avoids increasing the aerodynamic shape of the aircraft and does not encroach on usable passenger or payload areas of the aircraft," according to the patent. One drawing in the patent shows the tank's horizontal plane tilting downward toward the front to its lowest point, right behind and below cockpit level, presumably to accommodate doors in the front and access to the plane's interior. Compared to similarly sized tube-and-wing planes, a jet fuel–powered blended wing-body type plane has 50 percent more internal fuel volume than needed for a mission, the patent said: "Thus the incremental increase in fuel volume required for a blended wing body aircraft powered by liquid hydrogen is less than required for conventional configurations." Boeing is not saying much about its work on blended wing-body planes but is reported to have been working on the concept for decades. Together with NASA, Boeing has flight-tested unmanned small-scale models—21 foot wingspan, almost 400 pounds weight, top speed 120 knots—of blended wing-body planes, dubbed X-48B, at Edwards Air Force Base in California.

Flying into Space: Hydrogen Aerospace Planes

With interest in hydrogen aircraft waning as the twentieth century drew to a close, some influential folks in Washington, D.C., and elsewhere were paying increasing attention to finding a cheaper-to-operate successor to the liquid-hydrogen-and-liquid-oxygen-fueled space shuttle. Conceived as a low-cost utility truck of the space age, the shuttle had turned out to be less utilitarian and much more costly to operate than its designers had anticipated.

Thus, the 1980s saw the beginning of a quest for an aerospace plane: instead of blasting off vertically from a launch pad as the shuttle did, this one would take off from a long conventional runway, go into orbit, reenter the atmosphere, and land like a conventional plane, to be refueled with liquid hydrogen and be ready to take off again with a shorter turn-around than that required by the shuttle and, it was hoped, at a fraction of the shuttle's cost.[17] Senator Matsunaga's 1988 allusion to a "trans-atmospheric aircraft" was a reference to what Ronald Reagan had called the "Orient Express" in his February 1986 State of the Union address ("We are going forward with research on a new Orient Express that could, by the end of the next decade, take off from Dulles Airport and accelerate up to 25 times the speed of sound, attaining low-Earth orbit or flying to Tokyo within two hours").

The so-called Orient Express never flew but the idea morphed into the military's national aerospace plane (NASP) project: a single-stage suborbital hydrogen-fueled airplane-rocket combination that could reach any point on the globe within a few of hours. Begun two months after Reagan's 1986 State of the Union address, the NASP project, a high-priority, high-visibility joint effort of NASA and the Air Force, was the largest aerospace-plane program to be initiated in the latter half of the 1980s. (Other such efforts were mounted by Britain, Germany, and the Soviet Union.)

NASP was to be used as an experimental vehicle to test high-risk technologies for Reagan's Strategic Defense Initiative, but also to ferry people and equipment to future space stations more economically than could be done with the space shuttle. It would take off like a conventional airplane. Hydrogen-fueled, air-breathing "scramjet" (supersonic combustion ram jet) engines would push it at speeds of up to Mach 25 to the almost-vacuum of near space, where rocket power would provide the final push needed to reach a space station.

For a while, NASP officials, anxious to find allies, courted the small international hydrogen community. For example, Robert Barthelemy, director of the NASP joint program office, brought more than a dozen NASA and Air Force scientists and NASP contractors to the 1990 World Hydrogen Energy Conference in Hawaii, where they presented papers on various aspects of possible NASP-derived hydrogen energy technology. Ultimately, however, the NASP program was killed in the series of budget crunches in the early 1990s after expenditures of about $2 billion. As far as is officially known, no NASP was ever built.[18]

In the mid-1980s, British Aerospace developed the concept of a high-speed HOTOL (horizontal takeoff and landing) craft, envisioned as a recoverable substitute for the throwaway rockets then used to place payloads into space. Like the NASP, it was to be partially air breathing; however, it would not have achieved the extreme speeds and altitudes of the NASP. At the end of the 1980s, Germany developed the Sänger concept. Unlike the single-stage HOTOL and NASP, the Sänger was to be a two-stage, two-vehicle machine. With a big, manned, air-breathing, hydrogen-fueled, turbo-ramjet-powered first stage, it would take off from conventional runways and accelerate almost to Mach 7, when it would launch a smaller vehicle into space, propelled by a rocket engine fueled by liquid hydrogen and liquid oxygen. Neither HOTOL nor Sänger was ever built.

Then a new effort to design a hypersonic transport passenger plane with global range got underway in Europe around 2005. Various European publications reported in the fall 2007 that the EU and the European Space Agency (ESA) had agreed to fund a second phase of an initial thirty-six-month, 7 million euro ($10 million) power plant study for a hydrogen-fueled hypersonic passenger plane labeled LAPCAT (for long–term advanced propulsion concepts and technologies) (figure 8.3). Studies for LAPCAT II, which got underway in October 2008 and are to be funded at 10.4 million euros ($ 13.8 million), including 7.4 million euros ($9.8 million) from the EU, address the concept of a hypersonic 400-ton, 300-passenger Mach 5 aircraft that would fly from Brussels to Sydney, Australia, in about 4.6 hours.

Although various designs are under study, the one that is getting most attention and is most prominently displayed on ESA's Web site is the LAPCAT A2 under development at the British company Reaction Engines, based in Oxfordshire, a company with long experience in this area. Its three founders and key executives, Alan Bond, Richard Varvill, and John Scott-Scott, had worked together on the HOTOL, for which funding was canceled in 1988. It would be a huge plane, bigger than anything flying commercially today: at 139 meters, it would be almost twice as long as

Figure 8.3
A rendering of a future hydrogen-fueled hypersonic LAPCAT plane project, supported by the European Union and the European Space Agency. Such a plane could whisk 300 passengers from Brussels to Sydney, Australia, in two to four hours.

the biggest current passenger transport, the 79.75-meter twin-deck Airbus A380. But at a gross takeoff weight of 400 tons, it would weigh less than the heaviest A380, the 592-ton freighter version. About half of that weight—198 tons—would be liquid hydrogen to fuel the plane's four Scimitar engines, a design that Reaction Engine's Web site says is based on a combination of existing gas turbine and subsonic ramjet technology, as well as a precooler—a sophisticated heat exchanger—to transfer heat generated by the inrushing air to the hydrogen fuel.

According to its Web site, Reaction Engine estimates it would take thirteen years and 22.6 billion euros ($29.8 billion—2006 prices but mid-2010 exchange rates) to develop the plane to commercial reality. The price for each plane was estimated at 639 million euros ($844 million) assuming a production run of 100 planes—all of which, Reaction Engines acknowledges, is speculative: aside from developing the engine technology, no real work on the actual plane has yet started.

A second Reaction Engines project, the SKYLON reusable single-stage-to-orbit spaceplane—son of HOTOL, in a sense—is described as much more realistic and almost near-term for those thinking in terms of decades. Unlike the LAPCAT, which is powered by air-breathing engines, SKYLON is powered by two SABRE hybrid air-breathing/rocket engines fueled by liquid hydrogen and atmospheric air from takeoff up to speeds of Mach 5, switching to liquid oxygen at higher speeds up to orbital velocities in the vacuum of space. At 82 meters in length, it is smaller than LAPCAT, would weigh 41 tons empty, and would carry 66 tons of liquid hydrogen plus 150 tons of liquid oxygen. and would carry a 12-ton containerized payload up to a 300 kilometers equatorial orbit, or a cabin module for thirty passengers. Like LAPCAT, it would take off from a conventional but heavily reinforced, very long runway (5.5 kilometers) and would land there again as well. Another difference between the two craft is the onboard electrical power supply: SKYLON would use fuel cells as power sources, like the space shuttle, while LAPCAT would tap the engine, "like the airliner that it is," as Mark Hempsell, the Future Programmes director for Reaction Engines, put it in an e-mail.

"The prospects for SKYLON look very good," Hempsell added. "It is a commercial project with enough rate of return to attract investors and enough jobs to justify government support. SKYLON'S problems within Europe are more political as it is perceived as a counter to the vested interests of the expendable launch vehicle industry," presumably a reference to Europe's satellite launch company Arianespace, and perhaps Russian commercial launch activities. Reaction Engines expects to have SKYLON production prototypes flying for final testing around 2018 and ready to go into service with commercial operators in 2020.

As to a possible LAPCAT launch date, Hempsell said in 2010 that there is "no real answer at this stage but definitely well after SKYLON." But then in June 2011, hydrogen as aviation fuel—and as a spectacular example of launching advanced energy technologies generally—got a big boost with the announcement by European aerospace giant EADS, the parent of airplane manufacturer Airbus, that it is working on an environmentally benign hydrogen-fueled hypersonic transport plane that would cover the Tokyo-Los Angeles route in two and a half hours. EADS announced a concept study, apparently the first evidence of interest in hydrogen as commercial aviation fuel by a major planemaker in more than a decade, at a press conference on the second day of the annual Paris Air Show Le Bourget. EADS, which stands for European Aeronautic Defence and Space Company, said it expects to build demonstrators of the ZEHST (Zero Emission High Supersonic Transport) by the end of this decade, to be followed by development toward the goal of an

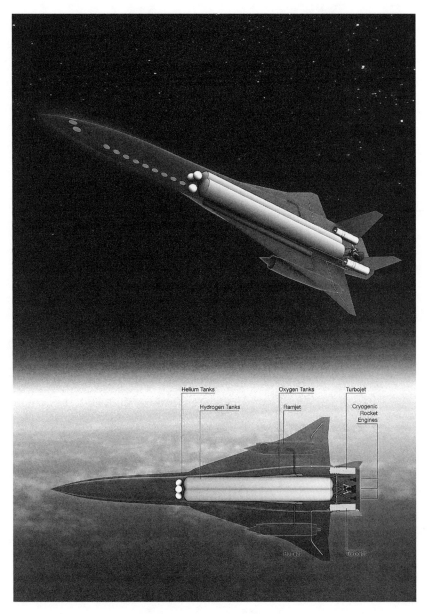

Figure 8.4
A rendering of the Zero Emission High Supersonic Transport plane concept
announced by EADS at the 2011 Paris Air Show.

operational 100-passenger Mach 4-plus vehicle. However, a commercial version will not rocket to its target operational altitude of about 32 kilometers (20 miles, or 105,000 feet) until three decades from now, said Jean Botti, EADS chief technical officer, at a press conference: "The initial concept represents a propulsion system architecture which is driven by flight safety considerations and by the requirements to minimize exhaust gas and noise emissions, in particular to mitigate the sonic boom," Botti was quoted in the announcement as saying. "We are on a very early stage with this research program. First series planes flying with this technology resulting from this concept will not fly before 2040." The plane will use hydrogen as its main fuel in its rocket engines, with an assist by an algae-derived biofuel feeding turbojet engines for takeoff and climb to about 5 kilometers before liquid hydrogen-fed rockets would take over.[19]

In its announcement EADS said that while reducing travel times for passengers is a key driver in designing future air transportation systems, these planes "will need to meet the air industry's ambitious environmental protection goals." The European Commission's roadmap, "Flightpath 2050," has set targets of reducing aircraft carbon dioxide emissions by 75 percent, a reduction of NOx by 90 percent, and noise level by 65 percent compared to year 2000 levels, according to the statement.

9
Hydrogen as Utility Gas: Hydricity, and the Invisible Flame

Launched in 2005, Tokyo's Fuel Cell Expo has rapidly evolved into the world's biggest showcase for fuel cell technology. The 2010 expo was no exception: some 2,300 professionals from sixty-six countries—from Argentina to Vietnam—showed up for the technical conference, close to 400 companies exhibited products and services, and more than 80,000 visitors jostled for three days at the beginning of March through the crowded aisles and exhibitors' booths. For a seasoned American operative like Robert Rose, the executive director of the Washington, D.C.-based U.S. Fuel Cell Council and the conference's lead-off speaker, there was no question that commercialization was getting underway in earnest. Although fuel cells are still subsidized and customers have to pay "quite a bit" for them, he said, it was evident this technology was "getting a toehold in some markets" and there has been "a notable uptick" in parts of Asia and Europe, he said in a later interview. He discerned "a sense of optimism" among suppliers to the educational market, for instance, in hallway conversations and casual encounters: "There were smiles on people's faces. No question, people are selling. People in the supply chain are perking up."

For Rose, "the most astounding news" was the announcement that the Korean government had committed itself to supporting the installation of 2 million residential fuel cell systems by 2020, as reported by Dal-Ryung Park, the principal researcher of the New Energy and Environment Team/R&D Center at Korea Gas Corporation (to put this into perspective: Korea's population stands at 50 million). In his presentation, Park said the government will pay for 80 percent of the system's cost, with another 10 percent available from local governments; these subsidies will decline to 50 percent in 2013 and 30 percent in 2017. As of

Figure 9.1
Crowds throng the aisles of Tokyo's 2010 Fuel Cell Expo, the world's largest fuel cell event.

2009, 210 PEM units made by three Korean manufacturers—GS Fuel-Cell, HyoSung, and FuelCell Power—had been installed under a demonstration program, Park's presentation said, with plans to distribute 10,000 units by 2012. The vision, according to another Park slide, was "to become one of the global leaders in fuel cells," with Korea aiming to take 20 percent of the global market in coming decades, with exports worth $26 billion and creating 560,000 jobs at 150 companies.[1]

Korea is not relying exclusively on its domestic industry to move ahead. In one of the biggest deals yet, a relative U.S. newcomer, seven-year-old ClearEdge Power in Menlo Park, California, with manufacturing facilities in Oregon, signed a three-year exclusive $40 million contract in 2010 to deliver more than 800 of its high-temperature hybrid proton exchange membrane (PEM) 5 kilowatt ClearEdge5 fuel cell systems to Korea's LS Industrial Systems as a baseload power-and-heat source running on natural gas or biogas. Aat the production end of the utility business, FuelCell Energy has been cooperating for years with POSCO Power, South Korea's leading independent power producer, in a technical alliance manufacturing molten carbonate fuel cell systems. Both companies signed a ten-year manufacturing and distribution agreement in 2007, followed a year later with an order for 25.6 megawatts of FuelCell Energy power plants and fuel cell modules, bringing the grand total ordered by POSCO Power at that time to 38.2 megawatts.

The U.S. Fuel Cell Council's Rose was also impressed with reports at the conference about the strides Japan was making popularizing fuel cell technologies. For example, ENE-FARM is a four-year subsidy program launched in 2009 to promote the installation of residential fuel cell cogeneration systems that had been in development since 1998, with a total of 6.07 billion yen ($65.6 million) in fiscal year 2009. Tokyo Gas, one of six companies participating in the program, had installed 796 units, including 520 precommercial models before ENE-FARM got underway, reported Koichi Aonuma, executive officer of that company, at the 2010 Fuel Cell Expo gathering. The strategy includes bringing the price down from about 2.0 to 2.5 million yen ($21,600–27,000) in 2009 to less than 0.4 million yen ($4,322) by 2025 while doubling the system's durability from about 40,000 hours to around 90,000 hours. "It's an interesting collaborative big marketing activity," said Rose. "It's exceptional. 70–80 percent of the Japanese people know what ENE-FARM is." He added, somewhat wistfully, "It's the kind of thing we cannot do in the States"

As one indication of Asian, and especially Japanese, leadership in this field, a Japanese market research firm, the Fuji-Keizai Group, said in a report announced in time for the conference that Japan's fuel cell market will expand perhaps as much as 99-fold in the next decade and a half, from 16.3 billion yen ($197.5 million) in 2009 to as much as 1,497 yen billion ($18.135 billion) in 2025 for cars and housing alone, according to a March 7, 2010, *Kyodo News* story: 990 billion yen ($11.9 billion) for cars and 507 billion yen ($6.1 billion) for housing). And just ahead of the conference, a British organization for building professionals, the Chartered Institute of Building (CIOB), said Japan plans to supply the United Kingdom and Germany with fuel cells to heat and power homes. Citing a BBC report, the story noted that more than 5,000 such systems, subsidized at half-price, were already operating in Japan; that companies such as Panasonic were in talks with EU governments to bring the devices to Europe; and that interest from the German, Korean, and U.K. governments was "intense."

In its 2010 Industry Review, released at the Tokyo conference, the U.K.-based fuel cell industry market intelligence group Fuel Cell Today (an organization for market-based intelligence on the fuel cell industry owned by the noble metals and specialty chemicals group Johnson Matthey) pretty much confirmed Rose's impressions of Asia's leadership. It noted that during the previous three years, Asia had gained a leading position in system shipments, accounting for about 50 percent of the total in 2009 while shipments from European manufacturers dropped between 2008 and 2009, mainly, *Fuel Cell Today* believed, as a result of the global economic crisis.

Several national residential fuel cell programs are underway in Europe as well, but they pale in comparison to the Asian action. Still, Fuel Cell Today describes Denmark's program as unique, in line with what the report says is that country's position as "one of the world's leading nations in terms of hydrogen and fuel cells." In this program, the uniform size of these fuel cells is 1 kilowatt, the principal difference being that it is designed to follow load demands for heat—usually regarded as a by-product to electricity generation—as the primary control parameter. Also, the program pits different technologies—low-temperature PEM, high-temperature PEM, and solid oxide fuel cells—against each other. Launched in 2008, ten units running are on natural gas or renewable hydrogen, and in the upcoming phase 3, another 100 were to be added, with stacks contributed by Danish companies Topsoe and IRD, the balance of plant components by Danfoss, and the Dantherm company acting as integrator.

In Germany, eighty-eight residential combined heat-and-power fuel cell systems (1 kilowatt electric, 2 kilowatts thermal) have been installed so far under the auspices of the country's national 86 million euro ($119.9 million) Callux field test, part of the National Innovation Program for Hydrogen and Fuel Cell Technology. Three manufacturers—Baxi Innotech, Hexis, and Vaillant—are expected to install about 800 systems by 2012, both low-temperature PEM systems (Baxi) and solid oxide (Vaillant and Hexis). Also in the plan are five utilities—EnBW, E.ON Ruhrgas, EW, MVV Energie, and VNG Verbundnetz Gas—all of this coordinated by the Center for Solar Energy and Hydrogen Research in Ulm in southern Germany.

In contrast, the United States does not have what one industry observer-participant described as a serious national residential fuel cell support program, although the federal government does offer a $1,000 per kilowatt tax credit. Sporadic efforts are underway in various states. California's Self Generation Initiative, created in 2001 and extended in 2009 by Governor Arnold Schwarzenegger through 2015, is open to anyone who buys electricity or gas from an investor-owned utility such as Pacific Gas & Electric, Southern California Edison, Southern California Gas, and San Diego Gas & Electric. When using nonrenewable fuel such as natural gas, the 5-kilowatt ClearEdge5 system qualifies for a $12,500 rebate check under this plan, for example. Helped by this program, ClearEdge has sold baseload residential combined heat-and-power (CHP) systems in six California locations, reports a database on the Web site of the nonprofit Fuel Cells 2000. In 2001, the U.S. Defense Department's Engineer Research and Development Center began a series of demonstration projects over four years at a number of military facilities. Limited efforts are underway in New York, where the New York State Energy Research and Development Authority in 2009 funded home demonstration projects by Plug Power with three GenSys systems.

These strides toward fuel cell commercialization for utility applications are the outcome of developments that began in earnest toward the end of the final quarter of the twentieth century in many places, companies, and laboratories. It is impossible to list them all here, but prominent examples at the producers' end included companies such as International Fuel Cells (now UTC Power), and its Connecticut competitor and neighbor, FuelCell Energy (née Energy Research Corporation), working on near-megawatt-sized phosphoric acid and molten carbonate fuels cells running on natural gas. Canada's Ballard Power Systems was working to extend its leadership in PEM-type transportation fuel cells to stationary and utility uses. Siemens-Westinghouse Power Corporation in Orlando, Florida, was working on solid oxide fuel cells, including combined cycle variants, together with German utilities RWE and EnBW as well as with the National Fuel Cell Research Center at the University of California at Irvine, and with Shell Hydrogen, using Shell-developed carbon dioxide sequestration technology, near Bergen, Norway. Others working on stationary solid oxide fuel cell plants included Japan's Tonen,

Sanyo Electric, Murata, and Mitsubishi and, in Germany, Dornier, at the time a subsidiary of the Daimler-Benz group.

Targeting the consumer end, many companies in North America, Europe, and Japan began at the end of the 1990s to develop small, stand-alone fuel cell systems and components for residential use, usually fueled by natural gas. They were driven in part by hopes that the utility market would be deregulated and that large, centralized utilities would be broken up; ten years later, some of them were defunct. A partial list included Avista Labs, H Power Corporation, Plug Power, Manhattan Scientifics, Nuvera, Delphi Northwest Power, DAIS-Analytic, ElectroChem, DCH Technology, Thermo-Electric, ZeTek, De Nora, Sulzer, Fuji Electric, and Matsushita.

This emergence of fuel cell technology for utility applications at both the producer and the consumer ends marked a fairly radical change in directions of how hydrogen would fit into utility operations. Before the turn of the century, hydrogen was generally regarded as replacement for and supplement to natural gas, with natural gas serving as a bridge to hydrogen, and hydrogen eventually to be used almost like natural gas with special types of burners or in flameless catalytic combustion—the "invisible flame" metaphor.

With the emergence of fuel cells as viable, almost-mainstream, technology, the "invisible flame" notion has been replaced largely by the hydricity concept (see chapter 1)—the idea that both are energy carriers interchangeable by electrochemistry, the yin and yang of energy: "polar or seemingly contrary forces are interconnected and interdependent in the natural world, and how they give rise to each other in turn," as one article describes it pretty accurately.[2]

It was not always thus. A few people are still pursuing the "invisible flame" path that was dominant in the twentieth century. For example, in the late 1960s, the Institute of Gas Technology (IGT) exhibited a "Home for Tomorrow" said to be powered by "reformed natural gas," a "new super active form of natural gas." An illustrated four-page color brochure described new types of illumination, portable appliances, flameless wall-panel heaters, total climate control, and electricity generated in the house by means of fuel cells. The "super active form of natural gas" was simply a hydrogen-rich mixture, similar to the manufactured town gas or coal gas of the late nineteenth and early twentieth centuries.

Whereas town gas typically had up to 50 percent hydrogen, "reformed natural gas" would have contained about 80 percent hydrogen, 20 percent carbon dioxide (CO_2), and less than 0.5 percent CO.

The IGT brochure envisioned that straight natural gas would be piped to the home. Except for the portion that would be burned directly (e.g., for space heating), the gas would be piped into a compact "reformer" that, using heat and steam with the help of catalysts, would convert natural gas into "reformed natural gas"—mostly hydrogen. According to a 1970 article in the industry magazine *Appliance Engineer*, such a use of hydrogen would allow the use of catalytic burners on kitchen stoves. Unlike conventional burners, which burn with a blue flame, these burners' flames would be invisible.[3] They would transfer more usable heat. Temperature ranges would be much wider and could be adjusted much more finely, and the heat would be distributed much more evenly over the burner's surface. A catalytic burner, said the article, "is self-starting, requiring no pilot, glow coil or spark ignition." Ignition would take place when the hydrogen gas contacted the catalyst (typically a very thin coating of platinum) and the burner began to give off heat. The by-product would be water vapor. Because of the fine-tuning capability and the wide range of heat available, "water can be boiled very rapidly or the most delicate sauce kept barely warm," the *Appliance Engineer* article said. A warming tray could be made of wood because the temperature could be limited to a maximum 250°F, below wood's burning or scorching temperature. Room heaters could operate at low temperatures and could be hung on walls and covered with synthetic materials. "Reformed natural" gas was also to be the power source for candoluminescence, a novel illumination process in which light emitted by molecules, ions, and atoms would be stimulated by a flame. "The physical causes of candoluminescence are still unknown, but we do know that this phenomenon provides new product possibilities," the article said.

Though nothing commercially saleable came of these "Home of Tomorrow" pipe dreams, the research behind them provided a valuable first look at many of the technical problems that had to be solved. Undeterred, the IGT also began developing high-temperature catalytic gas burners. At the 1976 World Hydrogen Energy Conference, held in Miami Beach, IGT researchers Jon Pangborn, Maurice Scott, and John Sharer presented a paper on the technical problems of burning hydrogen in

modified conventional burners and in advanced catalytic burners. Pangborn, Scott, and Sharer said that hydrogen could not be directly substituted for natural gas in domestic and commercial appliances and that
burners would have to be modified, a prospect they did not regard as
daunting. "Similar equipment modifications were necessary when natural
gas was substituted for manufactured gas several decades ago in the
United States," they explained.

Conventional burners suffer from flashback when burning hydrogen.
Although the flame will not travel back into the pipe beyond the metering orifice (beyond that point, there is no air or oxygen to sustain the
flame), a flame traveling back to the orifice may damage the burner head.
The problem can be tackled by increasing the pressure of the gas and
using a smaller burner port or by reducing or eliminating the amount of
so-called primary air—air mixed into the gas stream near the metering
gas orifice.

Although hydrogen does not produce any carbon monoxide or any
unburned hydrocarbons, an open-flame hydrogen burner would produce
about 30 percent more nitrogen oxide emissions (due to the flame's reaction with nitrogen in the ambient air) than a comparable natural gas
burner. Pangborn, Scott, and Sharer felt that most gas appliances could
be made compatible with hydrogen by increasing the pressure drop and
the flow rate at the user's end and by redesigning the burner.

A few experimental hydrogen-powered houses have been built by
dedicated individuals and institutions. One of the earliest conversions
was undertaken by the late Olof Tegström in the mid-1980s. Primary
power for Tegström's 1,334-square-foot two-story home in Haernoesand,
on Sweden's east-central coast, was provided by a 72-foot-high windmill
with a three-blade propeller 49 feet in diameter. With two generators
(one for high and one for low wind speeds), the windmill generated about
40,000 kilowatt-hours per year. On average, 1 kilowatt was used to
power domestic appliances and 5 kilowatts for hydrogen generation. Hot
water was stored in a 5-cubic-meter tank for house heating. Tegström
also sold some electricity back to the grid. A modified kitchen stove with
a stainless steel grid made the hydrogen flame visible, generated almost
no nitric oxides, and provided humidification. In all, Tegström's highly
insulated house required about 20,000 kilowatt-hours per year.
Hydrogen was generated in an 84-volt solid polymer electrolyte (SPE)

electrolyzer that, at 50 amperes, produced 1 standard cubic meter of hydrogen per hour from 4 liters of tap water. After drying, the hydrogen was stored in a hydride tank, apparently made of a fairly standard iron-titanium alloy.

Tegström also converted his Saab 900 automobile to hydrogen power. He estimated that converting both the house and the car had cost the equivalent of about $139,000 (at average 1998 exchange rates). But "these outlays should be seen as a down payment for 20 years of energy costs," he said.

In the late 1980s, Walt Pyle began converting his 1,800-square-foot two-story house in Richmond, California, to photovoltaic and hydrogen energy. Pyle had started on his quest for clean power for his house in the mid-1970s by investigating photovoltaics and adding energy-efficiency improvements. Pyle, once a staff engineer at a major oil company and later a solar energy entrepreneur, is still at it; his 2010 home page lists several hydrogen products and appliances, including a hydrogen-fueled tabletop barbecue for $495 (plus shipping and handling—www.hionsolar.com/)

For primary power, Pyle relied principally on fifty-two photovoltaic panels producing about 1,500 watts peak. (For short-term storage of electricity produced by these panels, Pyle uses two strings of six industrial-type batteries with a total capacity of about 1,600 ampere-hours. He installed a small 1-kilowatt electrolyzer. He stores the hydrogen and oxygen in three converted medium-pressure tanks. For cooking, Pyle converted a conventional gas stove to run on hydrogen, modifying the burners to avoid any mixing with air before the hydrogen reaches the burner ports to avoid flashback.

For home heating, Pyle has converted four conventional catalytic multifuel wall-mounted space heaters, originally designed to burn natural gas and to be used in campers and recreational vehicles, to hydrogen or natural gas operation. Pyle stays off both the gas and the electricity grids for about nine months out of the year. He once estimated that he has spent about $30,000 to retrofit the house.

In 1990, the architect Markus Friedli switched his 2,367-square-foot home in the small Swiss town of Zollbrueck to solar and hydrogen energy and took it off the local electricity grid.[4] His four-year conversion program, assisted financially by the canton (province) of Berne, made use of commercially available components, including photovoltaic panels, an alkaline electrolyzer, and a hydride vessel. Also installed were a hydrogen purifier and a compressor. Friedli reported that he was using hydrogen to cook and run his washing machine. His hydrogen pipes were welded or screw-joined to gas-proof standards and were made of embrittlement-proof high-grade steel. Rooms containing hydrogen machinery were equipped with sensors able to detect hydrogen concentrations as low as 0.14 percent, set off alarms, and shut off the main hydrogen pipe. The total cost of the conversion at the time was said to be the equivalent of about $216,000. In 2010, Friedli sold his house after retirement, and the solar and hydrogen components, including a bifuel minivan that could run on either gasoline or hydrogen, were sold to an interested party in Holland.

The most ambitious hydrogen house in 1990s was built in Freiburg, Germany, designed and built by the Fraunhofer Institute for Solar Energy Systems. Planning for the $1.5 million, 1,566-square-foot, two-story experimental house (about $1 million for the sophisticated energy-efficient structure, the rest for solar and hydrogen energy technology) began in 1987. Hydrogen was used for long-term energy storage. Through the use of advanced foamed glass, transparent heat dam materials, and insulation made of reprocessed waste paper, heat losses were cut by more than 70 percent, and the house went "on sun" in 1990.

Visually, this grid-independent and furnaceless house was distinguished by a glass-covered curved facade facing south and the absence of a chimney. Mounted on the roof were 40 square meters of photovoltaic cells plus thermal solar collectors. A 2-kilowatt electrolyzer in the basement generated hydrogen for long-term storage for the winter. (A bank of lead-acid batteries storing about 20 kilowatt-hours was used for short-term solar electricity storage.) Hydrogen was reconverted into electricity by a fuel cell, and the specially designed four-burner catalytic stove cooked with hydrogen.[5] Hydrogen was stored as a gas in an aboveground tank alongside the house; it could hold the equivalent of 1,400 kilowatt-hours. Oxygen was stored in an underground tank. The project was stopped in 1995, with the house connected to the grid and turned into an office building.

Usually a gaseous fuel such as hydrogen is distributed by pipeline. One basic question is whether hydrogen could be carried in existing natural gas pipeline networks—something that many early researchers at one time seem almost to have taken for granted. But embrittlement is a continuing concern for both hydrogen storage in steel tanks and shipment by pipeline. Steel becomes brittle when it is exposed to hydrogen, and the higher the pressures and the temperatures, the more pronounced the problem is. One 1993 study noted that hydrogen generally can have a detrimental effect on toughness, ductility, burst strength and fatigue life and that pipeline steel is exposed to embrittlement already at normal temperatures. At the 1996 World Hydrogen Energy Conference, a number of reports dealt with materials and safety, and German and Japanese scientists presented papers that looked at hydrogen embrittlement of austenitic stainless steels. The German study concluded that higher nickel content made such steel more susceptible to hydrogen embrittlement. The Japanese study found that the ductility of certain weld metals decreased remarkably at cryogenic pressures, but that other steels were still sufficiently strong in their mechanical properties to be used in these harsh environments.

One method of preventing embrittlement cracks in future hydrogen pipelines may be the addition of very small amounts of other chemicals, such as oxygen or carbon monoxide. It has also been suggested that other additives should be mixed in as a safety measure that would either give a characteristic smell to normally odorless hydrogen or add a distinctive color to the normally invisible hydrogen flame—a warning to the user that hydrogen is around. This would be fine as long as hydrogen were to be burned directly in gas turbines or internal combustion engines—impurities generally do not interfere with the burning process—but may be problematical with fuel cells and their electrochemical reactions.

In their 1976 report, Pangborn, Scott, and Sharer said that before large stretches of pipeline could be converted to carry hydrogen, a great deal of research was required. "Any statement that hydrogen or hydrogen-rich gases can be adequately and safely delivered to the customer by using the in-place natural gas distribution system is a presumption," they wrote. There was not a single natural gas system but "a tremendous diversity of pipes and fittings." Early low-pressure manufactured-gas systems were made predominantly of cast iron. Since then, ductile and wrought iron mains have been added, followed by steel mains and service lines to the individual customer with the advent of higher-pressure natural gas. Utilities overwhelmingly use plastic pipes for new and replacement mains and services lines, the authors said. Then there are materials such as brass (used in valves), natural and synthetic rubber (used in mechanical joint seals and meter diaphragms), lead and jute (used as sealer materials), and cast aluminum (used in meter housings and regulator parts). It will have to be verified that these materials can be used safely with hydrogen.

Pangborn's team did not exactly dismiss embrittlement; however, they did not regard it as prohibitive either: "These modes of hydrogen attack have been observed to occur under pressure and temperature conditions far more extreme than those of a gas distribution system. . . . These metallurgical effects would not be expected to occur in distribution equipment used for hydrogen service because of the operating pressures and ambient temperatures."

Leakage, which some researchers considered a big challenge, was not that problematic, they asserted. In terms of volume, hydrogen loss is about two and a half times the rate of natural gas, but the energy loss is about the same as with natural gas. Plastic pipes are probably more permeable, but the researchers felt that the amount of hydrogen that would seep through the plastic would be "insignificant."

This report made no reference to high pressures, deemed essential by other researchers for efficient energy transport; the authors merely assumed that hydrogen gas would flow faster and at somewhat higher pressures. If existing pipeline materials were found to be safe for hydrogen, they suggested, then if the flow rate were simply increased to about 2.8 times the rate of natural gas but at standard, safe pipeline pressures, existing lines could deliver as much as 85 percent of the energy transported today by natural gas to the end user.

This was not to say that hydrogen is easy to use in every respect. Hydrogen does create greater problems than natural gas when new lines have to be installed or when old lines have to be repaired—during welding, for instance. Also, because hydrogen burns in such a wide range of air mixture ratios (from 4 to 74 percent hydrogen), new lines must be carefully purged with an inert gas—"an area of serious concern."[6]

The embrittlement issue is still around today, hovering in the background of current discussions of hydrogen infrastructure—the proverbial elephant in the room. An August 4, 2010, article in the Internet edition of a British publication, the *Engineer*, reported that researchers at the Fraunhofer Institute for Mechanics of Materials in Freiburg are studying hydrogen-induced embrittlement in a new special laboratory because it could affect everything from fuel tanks and parts of the fuel cell to ordinary components such as ball bearings.[7] The researchers are attempting to visualize what happens to metal components when exposed to hydrogen, and, said an August 2010 institute release, the team is using atomic and finite element method simulation to investigate the interaction between hydrogen and metal on both atomic and macroscopic scales. Embrittlement is something "not very well understood," the article quoted Fraunhofer researcher Nicholas Winzer as saying, despite the fact that the phenomenon has been known and written about since the early 1800s.

In the United States, three researchers from the University of Illinois at Urbana-Champaign reported on their work, which takes a combined materials science/mechanics approach to hydrogen embrittlement in pipeline steels, at the Department of Energy (DoE) annual Hydrogen Program Review in June 2010. Petros Sofronis, Ian M. Robertson, and D. D. Johnson said their six-year, $1.5 million study due to end in 2011, undertaken with two national laboratories, Japan Automotive Industry, and five industrial companies, was started to come up with a mechanistic understanding of the problem in order to devise fracture criteria for safe and reliable pipeline operations under hydrogen pressures of at least 15 megapascals (2,175 psi) and both static and cyclic loading conditions. "Such fracture criteria are lacking, and there are no codes and standards for reliable and safe operation of pipelines in the presence of hydrogen," the three said in their presentation. "Current design guidelines for pipelines only tacitly address subcritical cracking by applying arbitrary and conservative safety factors on the applied stress."

As renewable energy—wind and solar—gains traction as a future energy mainstay, replacing coal and natural gas in utility power generation, the idea of storing very large amounts of hydrogen gas in underground caverns to smooth out renewables' intermittency to ensure a steady supply of electric power has become the subject of serious study. The German electrical and equipment giant Siemens, for example, is developing end-to-end solutions for generating and storing huge quantities of hydrogen with megawatt-sized PEM pressure electrolyzers in large, underground salt caverns. In its research labs in St. Petersburg, Russia, Siemens is also investigating hydrogen-fueled big advanced gas turbines that are able to handle the higher temperatures encountered when combusting hydrogen.[8] Siemens is not particularly secretive about it either: articles about this work appeared in the fall 2009 issue of the Siemens' semiannual technology magazine, *Pictures of the Future,* but as of summer 2011, the company had not released any specifics such as performance, timetables, and locations of possible demonstration caverns and hydrogen turbines.

In 2010, gas turbine power plants that would combust hydrogen were also being developed as part of two carbon capture-and-sequestration (CCS) projects launched by a joint BP/Rio Tinto subsidiary, Hydrogen Energy, in California and Abu Dhabi. The Hydrogen Energy California project includes installation of a 250-megawatt (net) integrated gasification combined cycle turbine plant near Bakersfield in Kern County fueled by hydrogen extracted from petroleum coke, a refinery by-product. In Abu Dhabi, a 420-megawatt gas turbine is to be installed, part of the emirate's ambitious $15 billion all-green Masdar Initiative that will burn hydrogen extracted from natural gas.

Salt cavern hydrogen storage is intriguing carmakers. Charlie Freese, executive director at General Motors for global fuel cell activities, told a reporter shortly after his appointment in 2009, in order to meet the stated goal of cutting CO_2 emissions about 80 percent by 2050, fuel cells and hydrogen are critical components: "The only real high-density energy storage device to store energy on a scale of what some of these new renewable energy production facilities would require is really going to be compressed hydrogen," he said.[9] At the 2009 National Hydrogen Association annual meeting in Columbia, South Carolina, Freese explained by comparing ways of storing energy in a 2 million cubic meter salt cavern. With compressed air storage, the cavern could hold about 4,000 megawatt-hours, providing electric energy to recharge, say, 1 million GM Volt electrics or run a central European national grid "for some minutes or hours," according to his slide. Filling the same-size cavern with compressed hydrogen could store 600,000 megawatt-hours of energy, equal to 3.6 million tank fills for a fuel cell car or providing a grid storage buffer "for several days."[10]

Hydrogen storage caverns exist already. One, Clement Dome with a volume of 580,000 cubic meters, is near Mont Belvieu, Texas; another, Moss Bluff, is near Deer Park, Texas. Both are part of the pet-rochemical hydrogen grid stretching across southern Texas to near Lake Charles, Louisiana, according to presentations by a German underground storage company, KBB Underground Technologies.[11] In Europe, three 70,000-cubic-meter hydrogen caverns owned by Sabic (Saudi Basic Industries Corporation) Petrochemicals in Teesside in the United Kingdom store the gas at 45 bar (650 psi). Fifty years ago, hydrogen-rich town gas was stored underground in an aquifer some 20 miles west of Paris, near the small town of Beynes. In the oil-shocked hydrogen community of the 1970s, Beynes carried a certain amount of name recognition for a while as a successful example of hydrogen storage. Ironically, this facility came to the attention of the hydrogen community only after French authorities were making plans to phase out the town gas and convert the facility to store natural gas in the mid- and late 1960s.

At the 2010 World Hydrogen Energy Conference in Essen, Germany, a paper by four researchers discussed large-volume underground hydro-gen storage in some detail. The report, "Large-Scale Hydrogen Under-ground Storage for Securing Future Energy Supplies," said that if all of Europe were supplied with energy solely from wind and solar, the only feasible way to store energy and smooth out fluctuations would be hydrogen: 0.41 cubic kilometers of, typically, salt caverns to hold 167 terawatt-hours worth of hydrogen. Because of differences in energy, other methods would require less energy storage capacity but much larger volumes: pumped hydro, for a long time regarded as the utility storage method of choice, would need only 74 terawatt-hours but would require 106 square kilometers—about twice the volume of Lake Constance in southern Germany. A variant of compressed air energy storage, adiabatic CAES, would require 80 terawatt-hours and a storage volume of 2 cubic kilometers. The authors, Ulrich Buenger and Hubert Landinger of Munich's Ludwig-Bölkow-Systemtechnik (LBST), and Fritz Crotogino and Sabine Donadei, of KBB Underground Technologies, say the differ-ences are due to the fact that hydrogen energy storage is based on chemi-cal characteristics while the other two are based on physical characteristics. They explained that salt domes have been used for decades to store gases—primarily natural gas, but also others—because they are extremely gas tight, with theoretical leakage rates around 0.01 percent annually, and impervious to chemical reactions with hydrogen, with 20 megapascals (2,900 pounds per square inch) maximum operating pressure—about a third of the 10,000 pounds per square inch pressure in compressed hydrogen fuel tanks in current-generation fuel cell cars. In Germany alone, about 170 caverns are in use to store natural gas for seasonal load balancing, shut-down, extreme weather, and trading reserves. The main disadvantage is the relatively low round-trip efficiency—electricity to hydrogen and back to hydrogen—of less than 40 percent, but the authors said that hydrogen is nevertheless the only storage option to permit energy storage in these huge volumes. Because of the thick walls in a typical cavern—tens of meters to 100 meters—and the pressure from surrounding geological formations, gas caverns basically can never explode, asserted the authors.

Cryogenic liquid storage is the space-age approach to storing large amounts of hydrogen. In the 1980s and 1990s there was a certain amount of interest in it for utility applications, but that has faded away. As noted, a Hamburg utility tested storage of liquid hydrogen in conjunction with the operation of a stationary fuel cell. Cryogenic storage has some technical advantages. A 1972 American Gas Association report, "A Hydrogen Energy System," pointed out, for example, that at normal pressure, liquid hydrogen takes up only $\frac{1}{850}$ as much volume as gaseous hydrogen. Liquefying hydrogen is expensive, though. Commercial liquefaction processes consume about a third of hydrogen's energy content. There is promise for more efficient, less expensive refrigeration techniques with active magnetic regenerative liquefaction, a line of research pursued also in connection with work on superconducting magnets. John Barclay, now chief technology officer of Prometheus Energy Company in Redmond, Washington, who has been working in this area for more than three decades, presented the latest results of his ongoing work at the U.S. DoE Hydrogen Program's 2010 Annual Merit Review. Liquid hydrogen has been used and stored in fairly large quantities (typically between 15,000 and 26,000 gallons and in huge tanks in the 500,000-gallon range by industrial gas producers) for industrial use, but not as a fuel. The storage tanks at the Kennedy Space Center are much bigger—850,000 gallons and 900,000 gallons, respectively, for liquid hydrogen and liquid oxygen. Because of liquid hydrogen's peculiar characteristics, those containers were expensive, costing between $2 and $4 per gallon of storage capacity, according to the 1975 IGT survey. More recent estimates for very large liquid hydrogen tanks put the cost in the range of $10 to $12 million for a 1-million-gallon tank—2.5 to 6.0 times as much as in the 1970s.[12] They are dewars—complicated double-walled structures, similar to a thermos bottle in concept but with a stainless steel or aluminum inner liner, a perlite-filled vacuum between the two shells, and a steel outer casing. They are spherical because with that shape, the total surface is smallest in comparison to the volume of liquid stored, keeping evaporation losses to a minimum, and curved steel sections withstand atmospheric pressure better than straight walls do. Loss rates of about 0.1 per day for big tanks have been achieved. Today liquid oxygen and nitrogen tanks are flat-bottomed and cylindrically shaped because they are much cheaper to build than spherical ones. Some liquid hydrogen is always left in these tanks, not to build up pressure but to keep them permanently chilled; once they are allowed to heat up to ambient temperature, sizable quantities of expensive liquid hydrogen are needed to get the temperature down to storage levels again. The tanks are slightly pressurized to keep outside air from coming in; the air would freeze immediately, clogging up valves and other passages. Frozen solid oxygen from the air would present a hazard if it came into contact with hydrogen.

Aside from space applications, liquid hydrogen has not been widely used as a fuel. The double-walled piping and related equipment used in aerospace applications have been far too expensive for down-to-earth energy use. Japan's ambitious 1990s WE-NET program of turning hydrogen into the world's principal energy currency included some very large pieces of equipment that were ahead of their time, including a 50,000-cubic-meter underground storage tank, development of large liquid-hydrogen tankers, and a 500 megawatt liquid hydrogen–burning gas turbine. That project died in March 1997.

In these early years, hydride storage of hydrogen, in which hydrogen is stored inside the metallic lattice structure of certain alloys such as titanium-iron, received attention in automotive applications and was a subject of intense interest for large-volume storage. Companies and institutions that investigated hydride storage for industrial- and utility-scale hydrogen in the 1970s included Allied Chemical, International Nickel,

Phillips Research, Sandia National Laboratory, Battelle Memorial Institute, Brookhaven National Laboratory, and Public Service Electric & Gas of Newark, New Jersey. Today the concept of large-scale hydrogen storage using hydrides has been just about discarded as awkward and material intensive. But in the 1970s, Brookhaven National Laboratory, for example, developed fairly detailed engineering concepts for a 26-megawatt electrical peak-power electric plant driven by hydrogen stored in large iron-titanium hydride storage beds. Hydride systems have also been proposed and investigated for hydrogen purification, compression without normal compressors, separation of hydrogen isotopes to produce deuterium for heavy water, heating, air conditioning, refrigeration, heat storage, and recovery of waste heat. Hydride hydrogen compressors would do away with conventional compression equipment by simply heating a saturated hydride tank. With different hydride combinations and multistage operations, pressures of up to 100 atmospheres are possible, using only low-grade heat as an energy source, according to an early paper by Frank Lynch and Ed Snape at the 1978 Zurich World Hydrogen Energy Conference.

Isotope separation, investigated at various times by Daimler-Benz, Brookhaven Lab, and General Electric, exploits the phenomenon that certain hydrides, such as titanium-nickel hydrides, preferentially absorb heavy hydrogen (deuterium).

Finally, the heat effects of hydrides—heat release during hydrogen absorption and heat take-up during hydrogen release, but at different levels and pressures for different alloys—open up prospects for heat storage, pumping, air conditioning, refrigeration, and power generation, Lynch and Snape reported. One of the first examples of such a system was the HYCSOS hydrogen conversion and storage system, which employed two hydride materials, lanthanum-nickel and calcium-nickel, developed experimentally by Argonne National Laboratory in the mid-1970s. By selectively shuttling hydrogen back and forth among four tanks, exploiting different temperature and pressure gradients of the hydride materials plus solar heat input, the desired effects of cooling or heating a room were achieved.

Newer concepts for developing hydride-powered air conditioners for buses and cars were developed in the decades that followed, but none of them made it to the commercial stage.

Cryo-adsorption, another early, and mostly untried, storage idea, was a compromise between cryogenic and hydride storage. Adsorption is a way of storing a gas or a liquid next to the surface of the adsorptive material, such as activated carbon or nickel silicate, rather than drawing it inside the storage medium (as in hydride storage). A cryogenic adsorption concept with temperatures around −320°F, the range used industrially in the liquefaction of nitrogen but nowhere as low as those used in pure cryogenic storage, was presented at the 1976 Miami Beach hydrogen conference. Unlike cryogenic storage, which proceeds at ambient pressure, cryo-adsorption requires fairly low pressure of about 60 bars (270 pounds per square inch).

Ruhr Revival

For most of the twentieth century, Germany's Ruhr region was synonymous with the country's industrial might—big steel plants, coal mines, big machinery, big guns. But in the late 1980s, that luster took on a tired brown tinge. The once-mighty Ruhr, like old-line steel and coal industries in many other countries, succumbed to the "rust belt" syndrome.

In that industrial territory along the banks of the Rhine are a number of large chemical plants producing basic chemicals that go into aspirin, food additives, dyes, fertilizers, plastics, fibers, and other products. A dozen of these plants have the distinction of being linked to the world's oldest, and most extensive, hydrogen pipeline network: a 143- mile system of buried steel pipelines wrapped in bitumen and plastic that traverses cities, crosses the Rhine in two places, and transports 20,000 pounds per hour in a 10-inch pipe at 290 pounds per square inch (20 bars), according to a 2007 Argonne National Laboratory report.

During the first burst of interest in hydrogen energy that followed the Arab oil shocks, the grid enjoyed something approaching celebrity status among advocates of hydrogen energy. They pointed to it as proof that hydrogen can be transported safely and economically like natural gas. A 2006 paper by the German Hydrogen and Fuel Cells Association said that 1,390 kilometers (868 miles) of hydrogen pipelines were operating in Europe, with pressures ranging from 0.5 megapascals (73 psi) megapascals 30 MPA (4,350 psi).

Citing U.S. Department of Transportation and Argonne National Laboratory estimates, a Wikipedia entry said that some 900 miles of low-pressure hydrogen pipelines were operating in the United States in 2004 – insignificant compared to the national natural gas transmission network, which, according to the Argonne report, totals 180,000 miles.

Today, the system that had its beginning in the late 1930s with a humble 14.3-mile three-point pipeline link is at the heart of a regional clean energy initiative by the state of North Rhine–Westphalia, the NRW Hydrogen HyWay—one of a growing number of regional hydrogen-centered efforts across Europe, in North America, in Asia, and elsewhere. Established in 2008 as part of a climate protection program set up by the state government and with additional funding from the EU, the goal is to advance the market readiness of hydrogen and fuel cell technologies by development and demonstration projects such as transit bus fleets including an 18-meter (60-foot) Dutch-German fuel cell-battery-hybrid bus and two fuel cell midi-buses; pool passenger cars and light-duty vehicles operated by businesses and government agencies; a few hydrogen fueling stations; a "virtual power plant" based on decentralized fuel cell installations in the Ruhr area; uninterrupted power applications; and special applications such as forklifts. Hydrogen HyWay is the outgrowth of an earlier North Rhine–Westphalia initiative launched in 2000, the Fuel Cell and Hydrogen Network, with 350 members from business and science and claimed to be the biggest of its kind in Europe.

In the 1960s and the 1970s, when nuclear power was still untarnished and in full bloom, Cesare Marchetti and other energy strategists argued that massive nuclear plants, which were then thought likely, would have to be located far from population centers. But beyond 1,000 miles or so, transporting energy via electricity would become disproportionately costly, they said; the costs of transporting energy via hydrogen would increase less drastically.

Although the basic feasibility of hydrogen pipelines is not in dispute, the details still are because of myriad variables that come into play, including pipeline diameters and pressures, spacing of compression stations, materials, embrittlement, fuel costs for pumping stations, types of compressors available, and the geographical locations of sources of hydrogen.

In the late 1980s, an indication of the size of the investments required for long-distance transport of energy—hydrogen, electricity, or both—was provided by a detailed theoretical analysis of shipping solar energy from North Africa to West Germany.[13] The study assumed conversion and transport of high-voltage direct current (800 kilovolts on land, 500 kilovolts underwater) by cables traversing the Strait of Messina, pipelining of gaseous hydrogen, and a combination of the two. The idea was to assess the cost of shipping this energy from Algeria to North Rhine–Westphalia. The conclusion was that a combination of piped hydrogen and transmitted high-voltage electricity provided the lowest transport cost because it offered the smallest transport and conversion losses: 18 percent versus 24 percent for pure electric transmission and 27 percent for piped hydrogen. The calculated investment costs for such a system would be enormous and probably prohibitive: the total for the combination system came to 245 billion deutsche marks (at 1988 prices—$144 billion at the autumn 1997 exchange rate of 1.70 deutsche marks per dollar). A "pure" hydrogen pipeline system to carry the entire load would have required investments of 275 billion deutsche marks ($161.8 billion). The high-voltage-only system would have been the most expensive—a staggering 343 billion deutsche marks ($201.8 billion).

Three decades later, with threats of global warming and energy security looming larger than ever before, the notion of importing solar energy from North Africa with its gargantuan price tag no longer looks look prohibitive. It is now being promoted in earnest by global banks and energy companies in the visionary Desertec project (chapter 5).

But hydrogen is not the carrier of choice to transfer vast amounts of energy from North Africa to Europe: it is high-voltage electricity.

10

Nonenergy Uses of Hydrogen: Metallic H₂, Biodegradable Plastics, and H₂ Tofu

"Dry water."

It sounds like an oxymoron, but it is not. It's the pithy headline on a press release put out by the American Chemical Society at its 240th meeting in Boston in summer 2010 that described a presentation by a British researcher, Ben Carter, on work at the University of Liverpool. Dry water could be commercially useful, said Carter; it resembles powdered sugar and absorbs more than three times as much carbon dioxide as ordinary water and silica as a hydrate, Carter, a member of a research team at that university, explained that the substance became known as "dry water" because it consists of 95 percent water and yet is a dry powder. Each powder particle contains a water droplet surrounded by modified silica, the stuff that makes up ordinary beach sand. The silica coating prevents water droplets from combining and turning back into a liquid. The result is a fine powder that can soak up gases, which chemically combine with the water molecules to form hydrates. One potential application has to do with the ability of dry water to speed up catalyzed reactions between hydrogen gas and maleic acid to produce succinic acid, a feedstock widely used to make drugs, food ingredients, and other consumer products. Another application, one that Carter focused on, was its ability to store copious amounts of carbon dioxide (CO_2). It can also store methane, which may make it useful for collecting and storing natural gas and also perhaps as a storage mechanism for storing methane as vehicle fuel for natural gas-fueled vehicles. However, "a great deal of work remains to be done before we could reach that stage," Carter said. Dry water was discovered in 1968 and drew attention for its potential use in cosmetics. Scientists at the University of Hull rediscovered it in 2006 in order to study its structure, and the team headed by Andrew

Cooper at the University of Liverpool has since expanded its range of potential applications.

In the last decades of the twentieth century when hydrogen became the hot new frontier not only of energy but also of stretching resources in an environmentally benign way, research into nonenergy uses of hydrogen and applications bloomed in many niches: from the mundane and well established such as the production of ammonia fertilizers and hydrogenation of fats in the chemical industry, upgrading of high-polluting fossil fuels such as oil shale and tars, heat treatment of metals, to more esoteric pursuits such as making proteins or biodegradable plastics that never saw the light of big commercial success and the making of metallic hydrogen as an extra-energetic fuel and, perhaps, as a room-temperature superconductor. A few of these efforts will be described in this chapter as a matter of historical interest; some of the more arcane ones have not been heard of recently and appear to have faded into oblivion.

A Hydrogen Space Gun

One of the most recent, audaciously interesting, perhaps Jules Verne–flavored, projects is a gigantic space gun suspended in the ocean that uses hydrogen as an accelerant gas to shoot projectiles into space as a low-cost method of launching cargo into earth orbit. The concept, announced in early 2010 by three former Lawrence Livermore National Laboratory scientists, calls for constructing a stupendous 1.1-kilometer (0.68-mile) sea-based tube to shoot cargo into space at a fraction of the cost of ferrying up stuff with conventional rockets. The three, John Hunter, Harry Cartland, and Rick Twogood, have set up Quicklaunch in San Diego with private funding. They first built a smaller version of such a gun in the 1990s for shooting projectiles into space for high-altitude hypersonic engine research dubbed SHARP (Super High Altitude Research Project), partially inspired by a similar device at Lawrence Livermore designed to create metallic hydrogen, which will be discussed later in this chapter. The Quicklauncher supergun, they say, could ultimately fire half-ton payloads into orbits at perhaps $250 a pound instead of the current ballpark cost of around $5,000 a pound. Hydrogen is not used as a fuel here. Rather, it serves as a kind of gaseous slingshot: some

18,000 kilograms of hydrogen gas are contained in a chamber at the bottom of the tube, where it is heated by combusting natural gas via a heat exchanger to 1,450°C (2,600°F), causing the hydrogen gas pressure to increase some 500 percent, from 2,600 to 15,000 pounds per square inch. The operators then open the valve, and the projectile shoots up the gun tube and out into the atmosphere at 13,000 miles per hour, with an additional small rocket motor kicking in when the vehicle reaches 450 kilometers (281 miles) altitude to achieve low-earth orbit. The gun's nozzle immediately closes after the projectile's exit with a hydrogen capture valve, capturing and conserving about 97 percent of the propellant hydrogen gas for reuse. The nozzle sticks out above the surface resting on a crane-equipped platform that hovers above the water surface, suspended from the 10-foot diameter tube muzzle via bearings and held in position by tension members. Another nearby maintenance platform, similar to an offshore oil drilling platform, houses liquid natural gas and hydrogen tanks and provides the vehicle assembly area. Hunter presented the concept at the Space Investment Summit in September 2009 in Boston (a video is at www.youtube.com/watch?v=1IXYsDdPvbo). The Quick-launcher would be suspended with a buoyant, ballasted support structure in the ocean somewhere near the equator, where the earth's rapid rotation would give an additional launch boost. The floating cannon would dip at an angle as low as 1,600 feet below the surface, controlled with ballast and thrusters to achieve desired angles of inclination and different directions.

Another unusual, not widely known application is the use of hydrogen to clean up nuclear waste. A small start-up company that specialized in high-purity advanced hydrogen generators and gas purification equipment, H2Gen Innovations of Alexandria, Virginia (chapter 1), sold seven of these generators, each capable of producing 113 kilograms of high-grade hydrogen per day from natural gas, to France's AREVA Group in 2007. AREVA, the French builder of nuclear power plants, was a partner in Uranium Disposition Services of Lexington, Kentucky, a group selected by the U.S. Department of Energy (DoE) to design and build plants in Paducah, Kentucky, and Portsmouth, Ohio, to convert the government's inventory of depleted uranium hexafluoride, a by-product of producing nuclear weapons, into triuranium octoxide, a uranium compound and one of the forms of yellowcake and one of the most stable forms of uranium: it is the form of uranium found in nature. There is a lot of the stuff around: DoE said in a 2002 announcement that the inventory of uranium hexafluoride totaled 704,000 tons at three facilities in Tennessee, Ohio, and Kentucky. H2Gen Innovations was founded in 2001 by Sandy Thomas, a former legislative assistant to Senator Tom Harkin (D, Iowa) and Frank Lomax, the company's principal investigator and co-inventor of the company's advanced technology, which was recognized by DoE's hydrogen program in 2007 with an award for technology innovation in developing and commercializing small-scale hydrogen production. But financing problems forced the company to shut down in 2009, first with the sale of its advanced pressure swing adsorption technology to Chicago Bridge & Iron, and then by selling its hydrogen generation module technology to two affiliates of France's Air Liquide industrial gas company.[1]

Hydrogen is a good chemical-reducing agent. In chemistry, reduction is essentially the reverse of oxidation, the process in which materials react with the air's oxygen to form another compound. Rust (iron oxide) is the product of the interaction of iron and atmospheric oxygen; it can be transformed into iron again by exposure to hydrogen. (In industrial iron production, ores are reduced to metallic iron in a smelter.)

In the international chemical industry, hydrogen has been for decades one of the most important chemical raw materials for the production of many organic materials. It plays a role in making ammonia-based fertilizers, refining crude oil into petroleum, and making methanol, the basis for resins, varnishes, plastics, solvents, and antifreezes.

In the oil industry, refineries and other plants routinely produce and consume hydrogen. Hydrogen is needed in hydrotreating processes, in which sulfur and other impurities are removed during distillation of crude oil, and its importance is growing. An overview paper on the industrial uses of hydrogen presented at the 1996 World Hydrogen Energy Conference noted that "due to the increased use of heavier crude oils, containing higher amounts of sulfur and nitrogen and to meet stringent emission standards, need for hydrogen is experiencing a very rapid growth in the petroleum refining industry." The authors of this paper, Ram Ramachandran, Raghu Menon, Raymond Morton, and Thomas Bailey, of BOC Gases, said that industrial use of hydrogen can be broadly divided into four categories:

- As a reactant in hydrogenation processes. Hydrogen is used to produce lower-molecular-weight compounds or to saturate compounds to crack hydrocarbons or to remove sulfur and nitrogen compounds.
- As an O_2 scavenge to chemically remove trace amounts of O_2 to prevent oxidation and corrosion.
- As a fuel in rocket engines and potentially in automobiles.
- As a coolant in electrical generators to take advantage of its unique physical properties. In catalytic cracking of crude oil, hydrogen is produced as a by-product. In hydrocracking, large oil molecules are broken down with hydrogen into fuel distillates that are blended into gasoline.

Hydrogen has many industrial uses . Hydrogenation of edible organic oils made from soybeans, fish, cotton seed, peanuts, corn, and coconuts slows their propensity to oxidize and turn rancid. The addition of hydrogen converts a liquid oil into a solid fat, such as margarine or shortening, but this has raised many health concerns: hydrogenation, or partial hydrogenation of vegetable oils and fats produces transfats, which, according to a Wikipedia "Trans Fat" entry, carries an elevated risk of coronary heart disease but may also increase the risk of Alzheimer's disease, diabetes, and obesity. A brochure by the Mayo Clinic warns, "Trans fat raises your 'bad' (LDL) cholesterol and lowers your 'good' (HDL) cholesterol."[2]

Inedible tallow and grease treated with hydrogen can be used to produce soap and animal feed. In the manufacture of polypropylene, a type of plastic, hydrogen is used to control the molecular weight of the polymer. In a more recent application, hydrogen is used in plastics recycling: the plastic materials are melted, and then the molten plastic is hydrogenated to crack it to produce lighter molecules that can be reused to produce polymers.

One of the most widespread uses of hydrogen is to make ammonia fertilizer by means of a high-pressure reaction between nitrogen and hydrogen. Ammonia production consumes about half of the roughly 50 million metric tons of hydrogen produced annually worldwide (chapter 4).

In metallurgy, hydrogen is used in the reduction stage in the production of nickel. In electronics, it is used in the epitaxial growth of polysilicon by wafer and circuit manufacturers.

Hydrogen, both in pure form and mixed with nitrogen, is used as an oxygen scavenger in metalworking. Hydrogen atmospheres are used in the heat treating of ferrous metals to change some of their characteristics—for example, to improve ductility and machining quality, relieve stress, harden, increase tensile strength, and change magnetic or electrical characteristics. A hydrogen-nitrogen atmosphere is used in annealing steel to make it more machine workable, make it amenable to cold rolling, and reduce stress while the metal is being shaped or welded. Bright annealing makes for a smooth, shiny surface, like that of stainless steel. A hydrogen-nitrogen atmosphere is used to reduce or prevent oxidation and annealing of nonferrous metals as well. A reducing atmosphere of relatively pure hydrogen is used in processing tungsten, producing molybdenum, and producing magnesium via electrolysis from magnesium chloride.

Burning hydrogen with oxygen produces very high temperatures for cutting glass and the cutting and high-temperature melting of quartz, and in float glass manufacture, where the glass typically floats on a tin bath, a mixture of 4 percent hydrogen in nitrogen is used to prevent oxidation of the molten tin.

In the manufacture of electronics components such as vacuum tubes and light bulbs, brazing, a process in which materials are heat-bonded, is carried out in a hydrogen atmosphere or in nonreactive gases such as argon or nitrogen to prevent oxidization.

Hydrogen's unique physical properties also have applications in the electric power industry. Hydrogen has the lowest viscosity among fluids, Ramachandran and his colleagues wrote, and therefore it is best suited to reduce friction in rotating power equipment. Gaseous hydrogen is used to cool large generators, motors, and frequency changers. Hydrogen has a greater thermal conductivity than normal air and therefore provides better cooling. A hydrogen cooling system is a closed circuit in which the gas is routed via heat exchangers through the generator shell and through the stator windings. (There is some leakage.)

In nuclear research, liquid hydrogen is used to fill bubble chambers to make the traces of subatomic particles visible and photographable. In the nuclear power industry, hydrogen is used in some stages of nuclear fuel processing.

Hydrogen for Clean Steel Production

Hydrogen's ability to reduce ores to their metallic states as an alternative to conventional smelter methods has attracted the attention of researchers for more than three decades. One of the earliest proponents of this method was Tokiaki Tanaka of Hokkaido University, a specialist in nonferrous metals who presented a detailed review of hydrogen's potential in ore reduction in the December 1975 issue of the *Journal of Metals*.[3] Taking his cue from work on thermochemical hydrogen production cycles under way at the time at the Euratom research center in Ispra, Italy, and elsewhere, Tanaka said that the chemical reactions in these cycles dovetail nicely with the processing of sulfur-containing sulfide ores. As Tanaka saw it, the development of these nonpolluting methods of metal production was environmentally more important than the production of hydrogen. He saw a particularly great potential benefit in the use of hydrogen in copper smelting, which now takes three or four energy-intensive steps. He concluded that "future developments in the field of hydrogen technology may . . . bring about a revolution in the field of metal extraction as we now know it."

Today steel is produced usually with coal or natural gas, or in electric arc furnaces. But the idea of using hydrogen to reduce the industry's sizable carbon dioxide emissions has been around for a couple of decades, and there are signs that Asian steelmakers are planning to move in that direction in the coming decade. Hydrogen's potential and its ability to help preserve Brazil's rain forests were pointed out in the early 1990s by Lutero Carmo de Lima, a researcher with the Department of Mechanical Engineering at Brazil's Federal University of Uberlandia. De Lima suggested that electrolytic hydrogen could replace much of the charcoal made from rain forest wood then being used to produce iron and steel in Brazil, where charcoal-based iron and steel production accounted for about 26 million tons out of a total annual output of about 40 million tons. According to de Lima, some 20 percent of Brazil's annual energy consumption came from the burning of wood. (The biggest shares of Brazil's energy consumption came from hydropower and imported oil, at about 30 percent each.) De Lima, at the time working at the University of Miami's Clean Energy Research Institute, asserted in a paper that construction of several large hydropower dams in Amazonia could provide low-cost electricity, which eventually could be tapped for hydrogen production.[4] Realistically, he did not expect hydrogen to be cost-competitive with fossil fuels for a few decades, and large-scale substitution of hydrogen for charcoal probably would not begin until about 2020.

In the early 1990s, the Euro-Quebec Hydro-Hydrogen Pilot Project (EQHHPP), a bravely idealistic, but in the end unsuccessful, five-year attempt to ship Canadian hydroelectricity converted to liquid hydrogen to Hamburg for reconversion to electricity, embarked on a project to launch hydrogen-powered steelmaking with a small pilot plant in Ireland. The project's key movers were Joachim Gretz (then an official at the Joint Research Center of the European Community in Ispra), Willy Korf (a German steel executive with a reputation in the industry as a maverick), Tom Doyle (a former official of the European Community), and Raymond Lyons (the director of Industrial Consultants International and a director of the Irish firm Kent Steel). Korf apparently had planned to introduce steel reduction using hydrogen technology in his plants, but he died in a plane crash before he had a chance to carry out that plan.

In a 1990 article, Gretz, Korf, and Lyons noted that in conventional steelmaking, each kilogram of molten steel releases about 2.2 kilograms of carbon dioxide into the atmosphere and that the international steel industry contributes about 10 percent "of all carbon dioxide emissions caused by man with the combustion of fossil fuels."[5] "Hydrogen," they suggested, "is an excellent and clean means of reduction that produces steam rather than CO_2, and it does not generate any additional impurities as does coke (especially sulfur)."

The overall goal of EQHHPP's project was to set up hydrogen-fueled steel plants in some portions of the former Soviet Union, Brazil, and Ireland. None ever made it beyond the brainstorming stage. Nonetheless, the idea is still alive. Barry Brook, director of climate science at the Environment Institute of the University of Adelaide, mentioned it in his "Brave New Climate" blog in June 2009. Headlined, "Steel yourself—a clear role for hydrogen," Brook wrote he was made aware of the idea by an engineer from one of Australia's largest steel manufacturers. With steel responsible for what Brook said was about 7 percent of global CO_2, he could see a great role for hydrogen produced in plants adjacent to high-temperature nuclear reactors, with reactors and hydrogen plants next to major blast furnaces. More concretely, hydrogen is under consideration by ULCOS (Ultra-Low Carbon Dioxide Steelmaking), a consortium of forty-six European companies, including all major EU steel companies, and organizations united in an R&D initiative to drastically cut CO_2 emissions. Jean-Pierre Birat, an executive with steelmaker ArcelorMittal and European ULCOS coordinator, discussed prospects for hydrogen in steelmaking in a paper he presented at the First International Slag Valorisation Symposium in April 2009 in Leuven, Belgium. Hydrogen is one of three major solution paths in a shift away from coal where carbon would be replaced by hydrogen reduction or by electrolysis of coal (the other two are carbon capture and sequestration and the use of sustainable biomass). "Hydrogen steelmaking will depend heavily on the availability of green hydrogen," he wrote. Carbon capture and sequestration has been identified as a powerful solution for a long time, but he noted that "the optimism which prevails in many policy-driven publications is overrated." Apparently it is far from clear which technology will emerge on top. Cooperative programs and commercial-size demonstrator programs are now under way, which may lead to deployment after 2020, Birat said.

One Asian steelmaker apparently has made up his mind to go with hydrogen: South Korea's POSCO steelmaker, the world's fourth largest, hopes to end all of its carbon emissions generated in steel production, but not for another twelve years or so. A November 27, 2009, Reuters story quoted Park Ki-hong, a POSCO executive vice president, as telling a green technology seminar, "Our ultimate goal is to develop steel production technology [that would not] emit any carbon dioxide." Added a senior POSCO spokesman, "We are currently studying the hydrogen-steelmaking process. We hope to get hydrogen gas from small or mid-sized nuclear reactors, which are also under study by us." POSCO accounts for 10 percent of South Korea's total CO_2 emissions, according to the story. The hydrogen would be used to capture oxygen from molten iron ore instead of the current practice of using carbon from coal, Reuters said.

In a different technological context, the use of hydrogen-rich gas in the direct reduction of fine iron ores is already an industrial reality. There are several industrial processes operating on all continents, with two, the Midrex and the Hyl processes, dominating. Midrex Technologies says on its Web site that more than sixty direct-reduction modules have been built worldwide since 1969, with worldwide production of more than 60 million tons in 2006. An advanced variant, a fluidized bed process dubbed the fine iron ore reduction method, developed by Exxon, is in use in Venezuela.

Atomic, Solid and Metallic Hydrogen

Atomic, solid, and metallic hydrogen have been researched and specu-lated about since the 1930s in the United States, the Soviet Union, Europe, and Japan. The work continues to this day, seemingly at a slower pace than in its pre-2000 heyday when interest in hydrogen was strong everywhere. These materials continue to have a "laboratory curiosity" status, similar to that of liquid hydrogen from the 1930s to the 1950s. Researchers continue to cogitate about the properties of these materials. Hydrogen in metallic form is believed to form the core of the planet Jupiter. No useful quantities have been produced, but if they could be manufactured on an industrial scale, they might find applications as ultrapowerful rocket fuels or as room-temperature superconductors.

Atomic hydrogen (H) exists in only infinitesimally small percentages in normal gaseous hydrogen, which is almost always molecular (H_2) because the bonds between hydrogen atoms are very strong—stronger than any other chemical bond. If the two hydrogen atoms are somehow separated, their natural tendency is to immediately recombine into the H_2 molecule. This strong attraction, and the energy implied by this bond, have persuaded the U.S .Air Force to try to break up the hydrogen mol-ecule as a means of storing energy. When recombining into molecular hydrogen, atomic hydrogen would yield about four times as much energy as the best currently available chemical rocket fuel, liquid hydrogen combusted with liquid oxygen.

Since the early 1970s, NASA's Lewis Research Center in Cleveland, renamed the John H. Glenn Research Center in 1999, has been working on methods of turning a percentage of gaseous molecular hydrogen into atomic hydrogen. Initially this effort was based on the research of a German scientist, Rudiger Hess, whose 1971 doctoral dissertation detailed his efforts in splitting the hydrogen molecule into two atoms by running the gas through a pair of electrodes. According to a 1974 article in *Aviation Week and Space Technology*, Lewis researchers cooled a mixture of molecular and atomic hydrogen down to 4° Kelvin (barely above absolute zero), and the mixture then condensed out on the walls of the experimental apparatus.[6] The entire setup functioned inside a superconducting magnet field that theoretically aligned the electrons in the hydrogen atoms in such a way that they could not recombine into molecules. The researchers believed that eventually a method might evolve by which the atomic-molecular hydrogen mixture could be stable, though very low temperatures would be necessary.

Walter Peschka, the scientist who had converted BMW sedans to run on liquid hydrogen (chapter 6), was among the first to investigate these phenomena. (Rudiger Hess was one of his students.) In December 1978, Peschka managed to store 2 grams of spin-aligned atomic hydrogen for several hours. He estimated the energy that could be stored and released by reverting to the molecular state to be ten to twenty times the amount available from current chemical fuels.

Another exotic variant is solid hydrogen. First predicted by Eugene Wigner and Hillard Huntington in the early 1940s, solid hydrogen was produced in microscopic quantities in 1989 by two scientists at the Carnegie Institution's Geophysical Laboratory, Ho-Kwang (David) Mao and Russell Hemley, by squeezing samples of gaseous hydrogen under ambient conditions. In the June 23, 1989, issue of *Science*, Mao and Hemley reported that they had subjected hydrogen to enormous pressures—more than 2.5 megabars (2.5 million atmospheres). On the basis of optical measurements of light transmission and reflectivity, they claimed to have produced a "semi-metal"—a substance that conducts electricity, but not as well as a "real" metal. Twenty years later, Hemley was still investigating: In the August 3, 2009, issue of the journal *Physics* and in *Physical Review Letters,* he and two colleagues, Timothy Strobel and Maddury Somayazulu, reported they had gotten closer to metallic hydrogen with a hydrogen-silicon compound, $SiH_4(H_2)_2$, that, as the EurekAlert release put it ever so cautiously, "could be helpful in the search for metallic and superconducting forms of hydrogen." The three said the high-pressure tests involved a mixture of SiH_4—silane, an industrial compound—and hydrogen, making it possible to produce "fairly dense forms of hydrogen that do become metals at more experimentally accessible pressures." At about 7.5 gigapascals (GPa) (74,019.245 atmospheres), the team produced that new material, $SiH_4(H_2)_2$, "in which hydrogen bonds are unusually weak and which may become a metal at higher pressures"—by their estimate, only about one-tenth the pressure needed to turn pure hydrogen into a metal.

Metallic hydrogen is created by pressures even higher than those needed for solid, crystalline hydrogen. When molecular solid hydrogen is subjected to pressure, the hydrogen molecules, normally spaced about ten times as far from each other as the two hydrogen atoms forming the molecule, move closer together. Eventually the molecules are so close together that the distance is about equal to the atoms' distance from each other inside each molecule. At that point, the atoms of different molecules begin to interact. If the pressure goes still higher, a metal eventually forms.

Solid "metallic" hydrogen occurs under extremely high pressures— maybe 1 million atmospheres and maybe more (Hemley, Strobel, and Somayazulu said 4 million atmospheres). Apparently it is not certain at precisely what pressure the change from mere solid to metallic hydrogen occurs. It may be that not only hydrogen turns metallic. Some researchers have theorized that any material—"beach sand, a plastic spoon, a mug of beer, even the air you breathe"—would "become a shiny metallic solid like aluminum or copper at very high pressure," Robert Hazen, a research scientist at the Carnegie Institution and a professor of earth sciences at George Mason University, has written.[7]

Theory postulates that metallic hydrogen may have the same energy content as spin-aligned atomic hydrogen—about 50,000 calories per gram (1.4 million calories per ounce)—but that because it is assumed to be fourteen to fifteen times denser, it would pack a bigger wallop in a smaller space.

In his book *The Alchemists*, Hazen noted that a debate raged for years among high-pressure physics experts about what exactly Mao and Hemley had produced . He asked, "Does the observed darkening above 2.5 megabars correspond to the metallization long predicted by scientists, or is it the sign of some other exotic material? When metallization is confirmed, will scientists be able to stabilize the metal at room temperature?" "What does seem certain," Hazen continued, "is that hydrogen, compressed to unimaginable pressures between the flat faces of diamonds, forms a substance that is quite unlike anything anyone has seen before." Metallic hydrogen could be "a superconductor at room temperatures," he wrote, echoing theories that had been circulating for decades. It could be the "ultimate electronic material." Theorists calculate that it also could be the most concentrated form of chemical energy. "As a rocket fuel," said Hazen, "it could store hundreds of times more thrust per pound than any other material; as an explosive it would be thirty-five times more destructive than TNT."[8]

Apparently the U.S. military establishment has pursued—and may still be pursuing—the production of metallic hydrogen. "There could be no better way to pack hydrogen atoms into a hydrogen bomb than in its dense metal form," Hazen noted. "Keenly aware of such vast potential for destruction, some scientists raced to produce metallic hydrogen with new dark urgency. They were driven not so much by curiosity about the unknown as by fear that others might exploit the substance first." But many unanswered questions remain: "Once made, will the metal persist after pressure is released? Could you hold a chunk of metallic hydrogen in your hand, or would it be unstable and revert immediately to the gas form? Only the scientists in high-pressure laboratories can provide the answers."[9]

In a 1989 telephone interview, Mao had told me that the solid hydrogen he and Hemley had produced had returned to the gaseous state after the pressure was removed. Other theorists had speculated that the material would remain stable or at least metastable. A former collaborator of Mao, Peter Bell, speculated that metallic hydrogen could be made stable by "hydriding" it—adding small amounts of another metal, such as palladium, to turn it into an alloy. Mao said in the interview that this is a possibility, depending on whether the material is "quenchable"—the technical term for arresting the material in its altered state.

A radically different approach to making metal hydrogen has been pursued for almost twenty years by scientists at Lawrence Livermore National Laboratory in California: firing projectiles from huge guns at liquid-hydrogen-filled containers to very briefly subject the hydrogen to extreme pressures. As reported in 1996, William Nellis and his colleagues used a 60-foot-long two-stage gun.[10] According to the *New York Times*, "For a fraction of a second . . . the impact converted the molecular liquid hydrogen" into an "excellent conductor [of electricity], presumably a metal."[11] Fittingly, hydrogen gas was used in the gun's propellant system. In the first 30-foot stage, ignited hot gunpowder gases shot a piston forward, which drove hydrogen gas ahead of it. The hydrogen gas reached a very high pressure before bursting a disk-type rupture valve at the end of that first tube. That high-pressure gas then shot into a much thinner second tube holding the plastic-and-metal projectile, which then slammed into the liquid-hydrogen-bearing target, having reached speeds of up to 9 miles per second (more than 32,000 miles per hour), generating a pressure of 1.4 million megabars.

One important aspect of the findings—that conversion to a metallic state occurred at pressures much lower than predicted—may help explain why Jupiter has a huge magnetic field. The *Times* quoted Nellis as saying it meant that the hydrogen in the planet's core becomes a metal much closer to the surface than previously believed and that the upper 10 percent of the planet is much more electrically conductive than astronomers and scientists had thought. According to Nellis, "Metallic liquid hydrogen at a relatively shallow depth in Jupiter would help account for Jupiter's gigantic magnetic field, some 20 times greater than the earth's."

These efforts to reach extremely high pressures began decades earlier at Cornell University, the University of Maryland, the University of Osaka, and the Institute of High Pressure Physics in Moscow. A brief 1975 account by the Moscow team, headed by L. F. Vereshchagin, said the Russians had created what they thought was metallic hydrogen at temperatures of 4.2°K and pressures of about 1 million atmospheres.

The Soviet work and the Japanese efforts, published almost simultaneously, were met with skepticism. Robert Hazen called both claims "spurious." A change in electric resistivity reported by the Russians, which they thought was a sign that they had produced a true metal, was met with disbelief in the West. It was suspected, Hazen wrote, "that the drop in resistance ascribed to metallization was simply a short in an electrical circuit that had been crushed and drastically deformed by pressure."[12]

Atomic hydrogen, though, is another story. It is not a solid but a gas and can be created without pressure by other esoteric techniques, such as glow discharge in conjunction with a strong magnetic field at extremely low temperatures or tritium decay ("the method of choice for most of the most recent experiments" at Lawrence Livermore National Laboratory, according to a 1993 overview paper by Bryan Palaszewski of the Lewis Research Center).[13] Other technologies employ electron or other high-energy beams, nanotechnology, or microlasers.

Atomic hydrogen is spin-aligned, and it does not recombine into molecules because of the opposing electrical forces. This hydrogen state also displays characteristics, such as certain magnetic effects, that are expected to occur in metallic hydrogen. Experiments in making, storing, and handling tiny amounts of atomic hydrogen, some of them funded by NASA and the U.S. Air Force, have been going on for decades. Laboratories in the former Soviet Union are also experimenting with this material.

Palaszewski's summary described the fiendishly difficult challenges of making these otherworldly materials in usable quantities, but also some of the promises of atomic hydrogen: "Modern experiments use nanogram samples of atomic hydrogen, whereas up to many hundred tons may be required for each launch from Earth to orbit," he wrote.

The energy stored in atomic hydrogen would be much higher than what is now available from conventional rocket fuels. Theoretically, the specific impulse generated by the recombination of atomic hydrogen into its molecular form, without any chemical combustion via oxygen, could range from 600 to 1,500 seconds of specific impulse.[14] (The specific impulse of the space shuttle, powered by liquid hydrogen and liquid oxygen, is 453 seconds.) Put another way, a monatomic hydrogen-powered rocket could carry from 14 to 600 percent more payload, or the rocket's gross liftoff weight could be reduced by 82.7 percent, according to Palaszewski.

Nine years later, in a 2002 conference paper, Palaszewski described the basics of such a rocket: Atomic propellants are composed of atomic species—hydrogen, but boron or carbon could also be used, he wrote—stored in cryogenic solid hydrogen particles.[15] These particles, measuring only a few millimeters, are stored in liquid helium to prevent the recombination of the atoms into molecules. Once the hydrogen is warmed and the atoms are allowed to recombine, the recombination energy heats the hydrogen and helium to high temperatures, and the resulting gases can be directed in a traditional converging-diverging nozzle to create thrust and, theoretically, high specific impulse. Using a 15 weight percent atomic hydrogen fuel, a rocket's gross lift-off weight could be reduced by 50 percent, but the actual design and construction of such a rocket system will be intimidating, he acknowledged a year later. In a 2005 paper, Palaszewski wrote that "future propulsion systems using atomic rocket propellants with solid hydrogen will likely require massive facilities for creating particles and many complex processes to trap atoms. Though the complexities seem daunting, the potential of these propellants is great, and the capacity for reducing vehicle lift-off weight and increasing payload capacity is theoretically unmatched. In some future vehicles and energy systems, atomic propellants in solid hydrogen may allow us to store and controllably release large quantities of energy, and allow the final human expansion into the solar system."[16]

Hyperbaric Hydrogen to Fight Cancer

One of the least known and largely forgotten potential nonenergy applications of hydrogen is in the fight against cancer. A pioneer in this field was the late William Fife, a professor of hyperbaric medicine at Texas A&M University. Fife, who had been investigating hyperbaric hydrogen (hydrogen under pressure as "hydrox," a breathing mixture of hydrogen in a safe, nonexplosive ratio such as 95 percent hydrogen and 5 percent oxygen) for more than twenty years, said in a mid-1990s summary overview that the study of the effects of hydrox on living organisms was in its infancy, with no more than twenty-five scientists active in France, Sweden, Canada, and the United States at that time. "It must be emphasized that the field of hyperbaric hydrox for medical application is such a virgin field that virtually no definitive work has been done," he continued. "Thus, although our laboratory has carried out several preliminary studies, few scientists, and no industrial organizations, have realized the potential for medical applications of this gas mixture."[17]

Aside from medical applications, there has been little interest in the use of hydrogen-oxygen gas mixtures and no other research in the United States, largely because of the perception of danger, Fife added. An exception is the diving industry, which had been considering hydrogen-oxygen gas mixtures for divers. The most active, best-financed program was one in Marseille, where the Comex company had been conducting extensive tests, some in open water, for possible commercial operations.[18]

In a rare instance of media attention to Fife's work, the September 22, 1975, issue of *Medical World News* carried this one-paragraph item, headlined "Hydrogen Kills Tumors":

An experimental treatment using hyperbaric hydrogen has caused regression of squamous [medical synonym for "scaly"] cell carcinomas in mice. Baylor College investigators Malcolm Dole and F. Ray Wilson and William P. Fife of Texas A&M University put three mice into a hyperbaric chamber (along with food and water) which was then flushed with a mixture of 97.5 hydrogen and 2.5 percent oxygen at a pressure of 8.28 atmospheres. After an initial ten-day exposure, the tumors turned black, some dropped off, some appeared shrunk at their base and to be in the process of being "pinched off." None of the tumor effects were noted in control mice kept at normal room temperatures. Continual remission of multiple carcinomas were shown in mice returned to the chamber for another six days, Further research will study the permanency of these results.

Dole, Wilson, and Fife, who subsequently published their basic findings in the October 10, 1975, issue of *Science*, were not quite sure what had happened. In a March 1976 letter, Fife said, "This work is very preliminary and we do not yet understand the mechanism, or, indeed, the full scope of its potential." One possibility, Fife reasoned, was that molecular hydrogen under pressure scavenges "free radicals which are well known to cause some forms of cancer." Free radicals are atoms or multiatom molecules that "possess at least one unpaired electron": the atom or molecule is not internally balanced in terms of its electron-proton ratio but has excess electrical binding energy. "One hypothesis," said Fife, "is that destruction of these free radicals makes it possible for natural but weak immune systems of the body to cause the malignancy to regress."

As to the perception of danger when mixing hydrogen and oxygen, Fife "categorically" stated in his mid-1990s summary paper that hydrox is safe to use. The Texas A&M Hyperbaric Laboratory had accumulated more than 6,000 hours of exposing animals and humans to hydrox, at pressures ranging from 7 atmospheres absolute (ata) to 31 ata, for durations of up to 125 hours of continuous exposure, "without a single serious problem and without any accident related to the use of hydrox." In a 1970s experiment, Fife himself had lived on a nonexplosive hydrogen-oxygen mixture at a simulated depth of 200 feet for three hours and at a simulated depth of 300 feet for two hours. "I cannot tell the difference between hydrogen and helium when breathing it, and I could notice no ill effects," he wrote.

Fife died in 2008 at the age of ninety. His daughter, Caroline E. Fife, who as a teenager had worked with her father as a lab assistant at Texas A&M, believes the field is dormant. "Nobody has looked at it in twenty years," she said in a brief 2010 telephone interview. "There has just been no interest in it," but neither had anybody repudiated her father's findings. Caroline Fife, now chief medical officer of Intellicure, a maker of computerized charting tools used in wound care, hyperbaric medicine, pain management, and other areas based in the Woodlands, Texas, and associate professor at the University of Texas Medical School at Houston, said Texas A&M's hyperbaric laboratory had been dismantled.

Hydrogen to make food and plastics: a possibility of a future hydrogen economy that first attracted scientific attention in the early 1970s is the concept of producing food—single-cell protein—from hydrogen, with the help of certain bacteria and solar or nuclear energy. The idea got its start partly because of NASA's interest in closed-cycle methods of producing sustenance for future astronauts from recycled basic materials during long deep-space missions. Producing some kind of meat substitute refinery-style offered the prospect of reducing worries about how to supply an exploding world population with staple primary proteins, a major international policy issue two or three decades ago.

NASA's food-in-space research got underway in the early 1960s, the basic idea being to use hydrogen, CO_2, and mineral salts produced by astronauts. Electricity generated by a spacecraft's solar cells would produce hydrogen and oxygen from onboard water supplies in a permanent cycle. Two researchers from the Battelle Memorial Institute, John Foster and John Litchfield, first presented their concept of a continuous-culture machine using hydrogen for protein production aboard a spaceship for deep-space flights in 1964 at a national meeting of the American Institute of Chemical Engineers in Pittsburgh. NASA abandoned the project in the 1970s, Litchfield said in a July 1997 telephone interview. Lower animals were not adversely affected, he said, but "primates, humans, definitely were." Studies by nutrition researchers at the University of California at Berkeley showed "pretty clearly" that the material was unsuitable for humans.

As early as 1970, the European hydrogen pioneer Cesare Marchetti sketched in a lecture at Cornell University how hydrogen and certain bacteria can be a link between primary energy and the mechanism of food synthesis:

Chlorophyll is the keystone of the process. Energy from light's photons is accumulated by this phosphor and transferred into ATP, adenosintriphosphate, the universal energy carrier in biological systems. But when an organism oxidizes an energetic substrate, be it sugar or hydrogen, the result is the same. With ATP, the function of chlorophyll is taken up by . . . enzymes. The privileged position of chlorophyll is given by the fact that it is coupled to a primary-primary source of energy, the sun.

Nuclear fission, Marchetti noted, could be substituted for solar energy (Today, many, perhaps most, alternative-energy supporters would stick exclusively with the sun):

If we find a link between the biosphere and this new source, chlorophyll and agriculture are going to lose their privileged position and the corresponding limitations are likely to fall. Hydrogen can be the link. A certain number of microorganisms are able to use hydrogen oxidation as a source of energy and thrive on a completely inorganic substrate. . . . The energy is used in a quite efficient way to synthesize all sorts of things necessary to build and run the biological machinery-proteins, vitamins, carbohydrates, and so on. The energy conversion efficiency . . . is quite high, 60 to 70 percent in the best cases, 50 percent in an easy routine.

Marchetti made the intellectual leap of relating the energy output of a primary power source—nuclear, solar, or something else—to human food requirements:

A man needs a caloric input of 2500–3000 Kcal per day, corresponding to a rounded mean power of 150 watts. Taking into account all the losses from nuclear energy to hydrogen, and from hydrogen to food synthesized by micro-organisms, to have 150 watts "at the mouth" one should count on roughly 500 watts at the reactor level. This means that a reactor designed to run a power station of current commercial size, say, 1000 Megawatt [electric] . . . and assuming a conversion efficiency in the plant of 40 percent, [this] could be the primary energy source to feed about five million people.

In Germany, a team headed by Hans Guenter Schlegel at the Institute for Microbiology of Göttingen University had been investigating the properties of *Hydrogenomonas* bacteria since the 1950s. Schlegel's team also looked into the idea of a closed-loop system for future long-term space missions. But when global food supplies and possible future global food shortages started to loom as international policy issues in the 1960s and the 1970s, Schlegel began to investigate these bacteria more intensively from that point of view. Starting with small-culture dishes, the Göttingen team produced protein in fermenter tanks up to 200 liters in size. Some of the researchers were concerned mostly with protein production; others were looking at genetic manipulation of the hydrogen-devouring strains in an effort to come up with more efficient mutants. Schlegel explained in a 1971 article:

Basically, molecular hydrogen is not an unusual source of energy for living beings. All aerobic organisms derive the energy necessary for the construction of their cell substance and to maintain their life functions from the reaction between hydrogen and oxygen. Man as well derives his metabolic energy through the slow combustion of hydrogen, in other words, from the so-called Knallgas [German for detonating a mixture of hydrogen and oxygen] reaction, although he is not being offered hydrogen in its gaseous state as nourishment, but rather as part of his foodstuffs, weakly bonded to carbon. Metabolic energy is not released through the combustion of the carbon but primarily through the oxidation of the hydrogen contained in foodstuffs. The product of burning hydrogen is water.[19]

What about the very idea of eating "synthetic" food, made by bacterial action? In fact, bacteria have been used for centuries to prepare many foodstuffs and medicines, and they are being ingested all the time as cheese or yogurt. Schlegel said that after removal of indigestible or even toxic components, protein quality is high, and the nutritive value is similar to that of the casein found in milk. The protein can be extracted from the bacterial cell mass and processed similarly to the way soybean protein is turned into synthetic meat, sold commercially in health food stores and elsewhere as textured vegetable protein," still offered commercially in 2010. In a sense, the idea of producing protein in a laboratory or a factory is an extension of existing industrial practices. Decades ago, British Petroleum set up plants for producing protein from crude oil; other companies too have developed processes using natural gas as a basis for protein. Certain strains of yeastlike microorganisms grow rapidly on an oil base. That discovery, which dates back to the early 1950s, had been exploited industrially to make single-cell protein for animal feed. Similar techniques have been developed to make biodegradable plastics. Techniques for making biologically produced, biodegradable plastics were discovered at the Pasteur Institute in Paris in the 1920s.

In the 1980s, the British chemical industry giant ICI developed a biologically produced biodegradable plastic, trade-named Biopol, that was made essentially by a fermentation process that employed strains of *Hydrogenomenas*. These bacteria readily transform gaseous mixtures of hydrogen and carbon dioxide and oxygen for use as a slow-release encapsulation material for pharmaceuticals, among other things. Biopol was never a resounding commercial success. Monsanto, to which the process had migrated via various corporate changes, dropped Biopol in 1998 because these materials cost about ten times as much as comparably performing petroleum-based (and not biodegradable) commodity plastics, and because biodegradability was not a "big, positive selling aspect," according to a Monsanto executive.

Some scientists have been intrigued by the prospect of using microbes to produce protein from inorganic materials, especially if the process were to be driven by the sun. The organisms would use hydrogen as an energy source, CO_2 as a carbon source, air as a source of oxygen, and simple minerals from fertilizers to produce a protein-rich, cellular, spongy, whitish material. The late Paul Weaver, a biochemist and microbiologist at the National Renewable Energy Laboratory had been investigating this as a sideline to his main research task of making hydrogen fuel in the DoE's Hydrogen Program. Weaver had been looking at economical hydrogen production from organic matter: biomass or coal, for instance, but also municipal solid or tree chips. Weaver figured that hydrogen fuel production could be made cheaper by coproducing a higher-value material such as biodegradable plastics or high-protein animal feed.

Basically Weaver gasified biomass to make a fuel gas—primarily hydrogen and carbon monoxide. In a second step, the gas would be cleaned up by bacterial action, converting the carbon monoxide component into additional hydrogen in what is known as a water-gas shift reaction. The bacteria used for this step are known collectively as photosynthetic bacteria. Although the bacteria are photosynthetic (that is, sustained by sunlight), they perform the shift reaction equally well in either light or darkness for periods of more than a year.

On sunny days, the bacteria grow photosynthetically on a portion of the hydrogen and convert it into new cell material: single-cell protein, vitamins, cofactors, essential amino acids, or other nutrients. (When nitrogen is kept out of the process, the bacteria cannot make protein, and about 80 percent of the new cell material comes in the form of tiny plastic-containing granules.) Bacterial protein is similar to tofu, Weaver said. "It's also similar to egg white in that protein essentially has no taste. Our taste buds do not respond to protein." Weaver has sampled it: "It mostly dissolves on your tongue. If it were cooked, it probably would be spongy, as are egg white or tofu when denatured." The bacteria are essentially 100 percent efficient in converting the fuel gas components into new cell materials. Microbial single-cell protein could have a great impact on world health if generated economically, Weaver believed: "The general goal of industries interested in this product is to produce a high-protein animal feed, which is by far the most expensive component of producing the animal protein that the developed world desires for human consumption." Not only do microbes produce about 65 percent of their cell mass as high-quality protein (Weaver said the amino acid composition is similar to the World Health Organization's "ideal protein" standard); in addition, they contain nearly all essential vitamins and neutriceuticals. German, English, and Japanese companies were the most interested, according to Weaver, and companies involved in aquaculture (fish farming) the most active. These were the areas in which bacteria-driven hydrogen conversion might find a market niche, Weaver believed at the time—not the large-scale production of protein to head off starvation in the developing world, as envisioned thirty years earlier. "Remember," Weaver noted, "the human need for protein is about 20 percent of caloric intake, the rest being carbohydrates and fats. Since single cell proteins (SCPs) are 65 percent protein, supplementation of human food with minor amounts of SCP can improve nutrition significantly. Likewise, the bulk of animal feed is still grasses and grain, but protein needs to be added. The human world is not short of calories. There is a world shortage of protein, however, and more than a billion people show the effects of sub-optimal nutrition, such as in kwashiorkor [a form of malnutrition caused by protein deficiency, especially in young children in the tropics]." This line of research ended with Weaver, and as far as I can determine, nobody else is investigating the production of biodegradable plastics or protein from hydrogen. Pin-Ching Maness, a principal scientist at NREL who worked with Weaver in these areas, said in 2010 that there has been work producing polyhydroxyalkanoate, a bioplastic material, from waste and cheaper feedstock such as glycerol, but not from hydrogen. "Hydrogen is simply too valuable for the production of a chemical, especially with such a small market size for biodegradable plastics," she wrote in an e-mail. "However, I do see the value of it once the petroleum economy is directed toward renewable fuels, but perhaps not right now. The science is still valuable."

Is Intelligent Life Out There? The Cosmic Waterhole

Hydrogen, as an indicator of the presence of water, plays a big role in the search for extraterrestrial life, universally known as SETI (search for extraterrestrial intelligence). In an October 1992 story about NASA's then newly computerized search for intelligent life in space, *Newsweek* noted that the ongoing search by the planetary researcher Paul Horowitz was focused on hydrogen's frequency of 1,420 megahertz. "This simplest and most abundant atom in the universe," *Newsweek* explained, "vibrates at a frequency of 1,420,405,751 cycles a second, a frequency that Horowitz says would make sense as a meeting place in the vast radio spectrum."[20]

NASA stopped funding the radio search as an economy measure in the early 1990s, but Horowitz and his team continue to scan the heavens with private funding. At the time of the appearance of the *Newsweek* article in 1992, the Megachannel Extra Terrestrial Assay (META), largely paid for by the filmmaker Steven Spielberg, was scanning a huge number of frequencies while sticking fairly close to that of hydrogen. After searching 60 trillion channels over five years, META found thirty-seven "candidate events," as one Web site put it with sober restraint. None of them was detected again.

In 1995, META became BETA (Billion-Channel ExtraTerrestrial Assay), a faster, more discriminating, automatic sweep of 1 billion channels at a time, covering the full 1,400- to 1,700-megahertz band of "water hole" frequencies. Every candidate channel is checked twice— first in the east, then in the west. If a channel looks promising, the antenna is programmed to break off its survey and lock onto the candidate signal, sweeping on and off the exact wavelength to capture other nearby frequencies.

In 2006, Horowitz and his team changed gears, switching from scanning the sky with radio waves to light waves with an optical telescope, the largest east of the Mississippi. Installed at Harvard's Oak Ridge Observatory site in the rural town of Harvard, Massachusetts, the $400,000, 72-inch (1.8-meter) instrument (cheap by SETI-hunting standards) is the world's first large telescope designed exclusively for laser flashes that might be sent by across deep space by advanced civilizations thousands of light-years away. Scientists think mega- and petawatt-sized lasers (1 petawatt is 1 billion megawatts or 1 quadrillion watts; a 1.25 petawatt laser has been built at Lawrence Livermore National Laboratory, for example), flashing at thousands of times the brightness of our sun even thousands of light-years away for a trillionth of a second, would be plainly artificial. They would be easier to spot against dim starlight than faint radio signals buried in cosmic radiation, Horowitz and his team—as well as other researchers in the United States and Australia, for example—reasoned. And Horowitz's "light bucket," as he calls it, is ultra-quick: it takes 1 billion measurements per second—about the equivalent of reading all the world's books in print today, once every second.

The SETI Institute in Mountain View, California, is pursuing similar goals, but on a grander scale. The institute's initial search, Project Phoenix, lasted from 1995 to 2004, using the world's three largest radio telescopes—in Australia; Green Bank, West Virginia; and Arecibo, Puerto Rico—and observing more than 800 stars over more than 11,000 hours. It looked at a wider spectrum of signals, between 1,000 and 3,000 megahertz, in very narrow (1 hertz) channel segments, or about 2 billion channels for each target star. Project Phoenix did not cover the whole sky; rather, it listened to space around "nearby" (no more than 240 light-years away) sunlike stars considered most likely to have evolved long-lived planets that may be capable of supporting life. A much more ambitious and more expensive project, the Allen Telescope Array (ATA), got underway in 2001 with initial funding of $11.5 million, plus another $13.5 million two years later from Microsoft cofounder Paul Allen; the total cost of this joint effort of the institute and the Radio Astronomy Laboratory at the University of California, Berkeley, stood at $50 million. The goal is to build an array of what ultimately will be 350 small, linked, synchronized radio telescope dishes on a half-square-mile area at Hat Creek Radio Observatory, 290 miles northeast of San Francisco. The first phase was completed in 2007 with the installation of forty-two working antennas. The purpose is to conduct both radio astronomical research and SETI research. For the latter, ATA will survey some 1 million stars, looking for what SETI describes dryly as "non-natural extraterrestrial signals" out to 1,000 light-years; and survey the 4×10^{10} billion stars of the inner Galactic Plane in the "water hole" frequency of 1,420 to 1,720 megahertz "for very powerful non-natural transmitters."

Nothing has been detected yet. "No ET signals were detected," says SETI's Web site in its "Project Phoenix Frequently Asked Questions" section. And when in 2010, one clicked on the highlighted question, "Have we found aliens?" at Harvard's BETA Web site, an all-black screen popped up, with tiny letters squeezed into the upper left corner: "No, not yet."

11

Safety: The Hindenburg Syndrome, or "Don't Paint Your Dirigible with Rocket Fuel"

If you plan to park your future hydrogen car inside your garage, it might be prudent to keep other stuff, such as bikes, shelves, old furniture and file boxes, that rusty ice box stashed in a corner—anything that changes the shape of the garage's interior space—outside.

Just in case.

This was one of several conclusions and suggested precautions consumers should be aware of, according to a report by SRI International, the independent, nonprofit research institute based in Menlo Park, California. In a paper, "Experimental Study of Hydrogen Release Accidents in a Vehicle Garage," presented at an international hydrogen safety conference in Ajaccio, Corsica, France in September 2009, a team of SRI scientists reported this was the first time that actual hydrogen release and ignition tests were conducted inside a blast-hardened single-car garage built at SRI International's 480-acre remote test site near Tracy, California, one of the largest privately owned explosive tests sites in the United States. Previous hydrogen gas leak and ignition tests either used simulations with helium or did modeling studies only, according to the authors. The team investigated how natural and mechanical venting systems affected the impact from hydrogen that was released at various rates and ignited. For the tests with natural ventilation, two tests were performed with a hydrogen release rate of 9 kilograms per hour at the refueling interface of a fuel cell vehicle—one test with an empty garage and one with a vehicle inside the garage—and a third test with a vehicle inside the garage and low release rate of 0.88 kilograms per hour. For tests involving mechanical ventilation in the garage, release rates ranged from a low of 1.6 kilograms per hour to a maximum of 6.7 kilograms per hour, and with ventilation rates ranging from 0.1 to 0.4 cubic meters per second. Aside from keeping the garage tidy, keeping extraneous stuff out of the garage is the smart thing to do as a safety precaution: The scientists found in both modeling efforts and actual hydrogen gas release and ignition tests that changes in the confined space's geometry could increase the hazards associated with a confined hydrogen conflagration. Also to be taken into account is the vehicle's internal geometry, the researchers said. One important finding was that when the flame propagated through the vehicle, the vehicle's internal geometry induced turbulence in the flame front, which caused the flame speed to accelerate, leading to a significant increase in the consequences associated with the deflagration. Generally all the scenarios that the SRI team examined found that mechanical ventilation reduced the impact of hydrogen combustion, and when the hydrogen release rate was less than 4.9 kilograms, overpressures generated by the deflagration did not represent any risk to people or property. The research was funded by two Japanese organizations, the New Energy and Industrial Technology Development Organization and the Institute of Applied Energy.

Global Safety Research

Research into hydrogen safety issues, which began decades ago as efforts by individuals and organizations scattered around the world, has evolved into coherent organizational structures that collaborate globally today. It has also evolved into specialties of increasing diversity and complexity. The Department of Energy's (DoE) 2010 Annual Merit Review, for example, included fifteen research presentations addressing such topics as a national codes and standards template, component standards research and development, and education for emerging fuel cell technologies to hydrogen safety sensors. DoE team leader Antonio Ruiz said in his introductory presentation, "Safety, Codes and Standards," that this subprogram aims to help develop and adopt codes and standards for hydrogen and fuel cell technologies, promote safe practices industrywide, and facilitate harmonization of domestic and international standards (it was funded in 2010 at $8.8 million; the 2011 request was for $9 million). Among a host of achievements, Ruiz cited the development of hydrogen release behavior data, harmonization of domestic and international fuel quality specifications, development of safety courses for researchers and code officials, and creation of permitting workshops for more than 250 code officials and of first responder training courses for more than 90 responders from eighteen states. A December 2009 international safety workshop for compressed natural gas and hydrogen vehicles in Washington, D.C., marked the start of ongoing cooperation efforts with Brazil, Canada, China, and India, including plans for the Fourth International Conference on Hydrogen Safety in September 2011 in San Francisco.

Aside from these research projects, DoE is offering safety information and tools to both the general public and specialists on its hydrogen program Web site (http://www.hydrogen.energy.gov/). Among the offerings is an online course, "Introduction to Hydrogen for Code Officials"; separately, an online course, "Introduction to Hydrogen Safety for First Responders," including a CD (DoE will not ship it to other countries); a bibliographic database; and the quarterly *H₂ Safety Snapshot* newsletter.

In Europe, the nonprofit International Association for Hydrogen Safety was set up in 2001 with financial help from the European Union "to contribute to the safe transition to a more sustainable development in Europe by facilitating the safe introduction of hydrogen as an energy carrier of the future." Membership is open to anyone, and fees range from 3,000 euros annually for full membership for large industry or research organization to 100 euros for associate membership for individuals. The seventeen founding member groups came from Germany, France, Norway, Britain, Spain, France, Greece, Italy, Sweden, Poland and Canada; two U.S. institutions, Sandia National Laboratories and Pacific National Laboratory, are members.

The Hydrogen Syndrome

In the past, the word *hydrogen* has often spelled disaster and doom for the technologically timid; for some it still does. The term *hydrogen syndrome* conjures up the images of the great hydrogen-buoyed *Hindenburg* airship going down in flames while attempting to land in New Jersey in 1937. And the term *hydrogen bomb* calls up visions of planetary war among nations, slaughtering hundreds of thousands and rubbing out entire cities in one gigantic flash. But more than thirty years ago, an early hydrogen supporter, Dan Brewer, employed a real-life technological tragedy, the worst disaster in the history of civil aviation, to make the point that hydrogen is basically a benign, forgiving fuel: the collision of two 747 jumbo jets that crashed into each other on a foggy runway on Tenerife in the Canary Islands on March 27, 1977, when a KLM plane struck a Pan American plane just as it was lifting off. Of the 644 passengers aboard the two planes, 583 perished. Brewer, the former manager of Lockheed's hydrogen program and at that time probably the U.S. aircraft industry's most outspoken advocate of liquid hydrogen as a fuel, maintained that the carnage could have been much lower if the two planes had been fueled with liquid hydrogen instead of conventional jet fuel. After analyzing information that became available shortly after the disaster, Brewer concluded that many of the deaths were probably due to the kerosene-fueled fire that raged for about ten and a half hours. "If both aircraft had been fueled with liquid hydrogen, there is a reasonable possibility that many lives could have been saved," Brewer told an audience of experts six weeks after the tragedy at a hydrogen symposium at the European Community's Joint Research Center in Ispra, Italy. Some twenty-five or thirty passengers probably would have been killed outright by the direct impact, Brewer reasoned. Fire presumably would have broken out, but because of hydrogen's different burn characteristics, many passengers could have been rescued. Brewer's arguments went as follows:

- The fuel-fed portion of the fire would have lasted only a minute or two because hydrogen is so volatile and because it is highly unlikely that both liquid hydrogen tanks in both aircraft would have been ruptured.
- The fire would have been confined to a relatively small area. The liquid hydrogen would have vaporized and dispersed before it could have spread widely.[1]
- Radiation, or radiated heat from the fire, would have been significantly less, so that only those persons and that part of the structure directly above or next to the flames would have been burned.
- The absence of smoke from the burning hydrogen fuel might have saved some lives. However, smoke from other incendiary material in the plane probably would have negated this advantage for hydrogen.

Whatever the merits of Brewer's ideas, one cannot help but compare the Tenerife disaster, which faded fairly quickly from the public's consciousness, with the *Hindenburg* dirigible accident of 1937, in which thirty-five passengers and crew members and one member of the ground crew died. The disaster received saturation coverage in the news media, and many advocates of hydrogen as a fuel fear that it made *hydrogen* a negatively charged word in the popular consciousness.

The *Hindenburg* did not explode, as is often believed; it burned. Furthermore, its sister ship, the *Graf Zeppelin*, made regular scheduled transatlantic crossings from 1928 until its retirement in 1937 without a mishap. And before the disaster, the *Hindenburg* had successfully completed ten round trips between the United States and Europe.

Although sabotage was initially suspected as the cause of the conflagration, many have believed for a long time it was more likely that electrostatic charges present in the atmosphere after more than an hour of thunderstorms and rain ignited the hydrogen that was being vented as the crew was trying to tie the big dirigible to its mooring tower.

It should also be remembered that the Hindenburg and its sister dirigibles were designed to be inflated with nonflammable helium. Hydrogen had to be substituted when the United States, the sole source, refused to sell helium to the Germans in the aftermath of World War I.

The Hindenburg: A Cellulose Fire, Flavored with Hydrogen

Almost exactly sixty years after the disaster, the widely respected NASA veteran Addison Bain uncovered impressive evidence that persuasively exonerated hydrogen as the primary cause of the disaster and showed that static electricity and the presence of highly inflammable materials in the airship's skin were to blame. Bain, the former manager of the hydrogen program at the Kennedy Space Center, presented the findings of his decade-long investigation in spring 1997 at the annual meeting of the National Hydrogen Association. His research had included exhaustive analyses of surviving fragments of the *Hindenburg*'s cotton-based covering performed in NASA's Materials Science Laboratories. His conclusions, first published in *Air and Space Magazine,* were reported in a large article in the science section of the *New York Times.*[2] They generated a cover story in *Popular Science* and several TV documentaries as well.[3] In his keynote address, Bain said he had discovered that the use of cotton fabric and a doping process that had involved aluminized cellulose acetate butyrate and iron oxide had combined to produce a material likely to burst into flames with a minimum of incendiary incentive. "The Space Shuttle's solid rocket boosters use powdered aluminum as the fuel and iron oxide as a burning-rate catalyst," Bain noted drily.

Static electricity in the air (witnesses reported seeing blue discharges on top and in the back of the ship, near the point where the flames first occurred, moments before the fire) most likely provided the spark that set off the conflagration, Bain reasoned. "Atmospheric and airship conditions at Lakehurst were conducive to formation of a significant electrostatic activity on the airship," he noted. "Hydrogen naturally contributed to the fire," said Bain. "It is a fuel, and fuels must be flammable. But the airship envelope was sufficiently combustible so that it could have burned even if the dirigible had been filled with an inert gas like helium. It was really a cellulose fire, flavored with some hydrogen." The clincher of his presentation was his final slide, which showed a U.S. Navy dirigible on fire in July 1956 at a Naval Air Station in Georgia, in the rain, with firefighting equipment nearby. To the casual observer, it looked very much like the *Hindenburg*, with the envelope burning rapidly. But this airship had been buoyed by inert, nonflammable helium, not hydrogen. "The moral of the story," Bain concluded, is "don't paint your airship with rocket fuel."

Testing Hydrogen for Safety

Some of Dan Brewer's ideas about the behavior of liquid hydrogen were confirmed by a 1994 experiment in which copious amounts of liquid hydrogen were intentionally spilled and ignited in a row of abandoned military barracks near Berlin that had once housed Soviet troops. Scientists from Germany's BAM (Bundesanstalt für Materialforschung und -prüfung—Federal Institute for Materials Research and Testing), the German branch of the Battelle Memorial Institute, and the Jülich Research Center spilled some 650 liters of liquid hydrogen and about 100 kilograms of liquid petroleum gas (LPG), a mixture of liquid propane and butane. Each fuel formed big puddles on the ground, a metal sheet, and a water surface in a lane between two barracks, according to a paper presented at the

1994 International Cryogenic Engineering Conference in Genoa. The authors said that the series of tests, conducted over four days and employing some twenty sensors for the hydrogen test alone and five video cameras, represented the first attempt to find out what would happen if such spills were to occur in a residential environment. Previous spill experiments had yielded important data about vaporization and cloud formation, the authors said, but most of them had been conducted "in an open area under conditions that are not all typical for a real accident."

What happened? Not much. Liquid hydrogen did not exhibit any striking differences from liquid natural gas (LNG), LPG, and other conventional fuels. "What we have demonstrated is that even in the case of an accident, liquid hydrogen is not more dangerous than these other fuels," the researchers reported. In fact, it may be even safer in some respects because it evaporates quickly instead of spreading on the ground. A BAM press release said it was virtually certain that "liquid hydrogen is not any more dangerous in such an accident scenario than the well-known propane." According to the release, "The hydrogen cloud that had been created rose very quickly and dispersed, while propane initially collected on the ground but was subsequently dispersed by air turbulence."

As early as 1956, Arthur D. Little and Lockheed ran tests to get an idea of what hazards existed in handling the large quantities of hydrogen that were expected to be used in the U.S. space program, and of what procedures would be required. The Arthur D. Little researchers spilled and ignited as much as 5,000 gallons of liquid hydrogen in an open space. There were no explosions.

In a 1967 paper, A. A. Du Pont, a researcher for the Garrett Corporation of Los Angeles, reported that the "use of hydrogen in normal aircraft operation can be as safe or safer than the use of fuel," although "questions arise as to the relative hazard of hydrogen fuel in case of a serious accident where the airplane is partially destroyed and the fuel is spilled."[4] Du Pont said it is "very difficult" to obtain an explosive air-hydrogen mixture even in a confined space, a claim that others have disputed. He also asserted that a hydrogen fire is "very much better from a safety standpoint." "A spill of cryogenic hydrogen," he continued, "boils furiously on contact with the relatively warm ground, so the area of spilled fuel is therefore confined whereas the spill of hydrocarbon fuel spreads along the ground and covers a considerable area."

In the mid-1970s, researchers at Wright-Patterson Air Force Base fired armor-piercing incendiary and fragment simulator bullets into aluminum containers lined with polystyrene plastic, some filled with liquid hydrogen and others with petroleum-based JP-4 jet fuel. They also simulated lightning strikes into the containers. These experiments, they reported, showed liquid hydrogen to be "more forgiving" than kerosene. As described by Jack Lippert at the 1976 Miami Beach hydrogen conference, the results indicated that the incendiary weapons ignited but did not detonate the liquid hydrogen. The hydrogen fire was "less severe" and "expired more quickly" than a comparable JP-4 fire, even though the total heat content of the hydrogen sample was twice that of the JP-4 sample.

Blasting away at the sample containers with the nonexplosive fragmentation bullets demonstrated that the liquid hydrogen experienced little "hydraulic ram effect" (an internal pressure wave building up in the liquid that could rip open the container or tank and cause damage to surrounding sheet metal). The nonexplosive bullets also did not ignite the liquid hydrogen, which simply poured out through the entry and exit holes.

When a similar container covered with polystyrene plastic and filled with JP-4 was shot at, the bullet caused overpressure that forced the kerosene through the lid and ripped a fairly large exit hole. Lippert attributed the difference in behavior to the fact that liquid hydrogen is only about one-tenth as dense as the JP-4. A bullet creates much more impact and shock wave in a high-density fluid than in a low-density one.

To simulate lightning strikes, the researchers used a 6-million-volt generator to shoot big arcs into the liquid-hydrogen containers. Again, there was ignition but no explosion. "The data indicate," Lippert concluded, "that the hazards associated with liquid hydrogen utilization in combat aircraft may be less severe than those with JP-4," and "therefore it is recommended that liquid hydrogen should not be disregarded as future alternate fuel for military as well as commercial aircraft."

Some hydrogen supporters feel that in view of the complexity of rocket technology, NASA's experience may not be very relevant to the everyday safety problems likely to be posed by a future civilian hydrogen economy. Still, it is noteworthy that NASA has used and handled stupendous amounts of liquid hydrogen, most of it hauled by barge and tanker trailer over hundreds of miles to Cape Canaveral and other sites.

In 1974, a paper still cited occasionally today that reviewed ninety-six hydrogen incidents and accidents, was presented by NASA researcher Paul Ordin at the Ninth Intersociety Energy Conversion Engineering Conference in San Francisco. Ordin reported that NASA's tanker trailers had hauled more than 16 million gallons of liquid hydrogen for the Apollo-Saturn program alone. He described seventeen mishaps involving tanker trailers carrying between 3,000 and 16,000 gallons. Twelve of these had occurred during offloading; five involving highway accidents. Taken together, the results seemed to indicate that liquid hydrogen was fairly forgiving in potentially high-risk situations.

Another departure point for assessing the safety of hydrogen was provided by the experience of the world's first long-distance pipeline for transporting hydrogen to various chemical plants, operated by Chemische Werke Huels (CWH) in Germany and in operation today (chapter 9). Christian Isting, a Huels executive, said in a paper delivered at a 1974 European energy symposium that "compressed hydrogen frequently ignites on expansion." The reasons for this are not clearly understood yet, but they probably have to do with electrostatic charges caused by dust particles in the air. Pressurized hydrogen would almost always ignite if vented through a seldom-used line, Isting reported, but ignition could generally be avoided if initially, only small amounts were sent through the venting line and amounts and flow speed were increased gradually. Large amounts can be vented to the atmosphere without ignition except during thunderstorms. "During the many years of operation of the integrated pipeline network of CWH," Isting concluded, "explosions have not occurred. Either the hydrogen will ignite immediately on escape or, if no ignition takes place, an explosive mixture cannot form near the ground because of its low density. Explosions or even detonations in the pipeline need not be feared as explosive gas-air mixtures cannot build up because of the pressure prevailing in the line."

Hydrogen's Properties

To understand the safety implications of hydrogen more fully, it is useful to look at hydrogen's physical properties:

- Liquid hydrogen is a very cold material, –423°F (–252.8°C). Contact with human body tissue can result in severe burns, destroying tissue almost like the burn from a flame.

- As a gas, hydrogen diffuses very quickly into other areas, such as air. It has a very low density, so it rises quickly through the air.

- Mixed with air, hydrogen burns over a much wider range than methane or gasoline, for instance. A mixture of as low as 4 percent of hydrogen in air and as high as 74 percent will burn. The corresponding ranges for methane are 5.3 to 15 percent, and for gasoline, it is a very narrow range, from 1.5 to 7.6 percent. Jet fuel flammability range is even narrower, from 0.8 to 5.6 percent.

- When confined in a completely enclosed space—a room or a tank—hydrogen can be detonated, that is, exploded over a wide range of concentrations, ranging from 18 to 59 percent (by volume) in air. Methane explodes only in concentrations ranging from 6.3 to 14.0 percent, and gasoline and jet fuel detonate in a range from 1.1 to 3.3 percent.

- It takes very little energy to ignite a hydrogen flame, about 20 microjoules. Methane requires about twelve times as much energy, about 290 microjoules, to set off a burn, and gasoline needs about 240 microjoules.[5] However, out in the open without confinement, it is almost impossible to bring hydrogen to an explosion with a spark heat or flame. Hydrogen-air mixtures can be detonated only with a suitable initiator, such as a heavy blasting cap.

- The hydrogen flame is almost invisible in daylight (unless some special colorant is added), and it travels much faster than a methane flame. A hydrogen flame shoots upward at a rate of 2.75 meters (9 feet) per second, while methane and gasoline burn much more slowly, at 0.37 meters (1.2 feet) per second.

- Unlike kerosene or gasoline flames, hydrogen flames radiate very little energy, so heat is not felt at a distance.[6]

In 1993, a more general safety study commissioned by Germany's parliament, the Bundestag, concluded that the risks posed by the wide use of hydrogen are likely to be relatively low and could be managed fairly easily. "The technical risks in all components of a hydrogen energy system, from production to utilization, are, in principle, regarded as controllable," said the five-page executive summary.

One of the best-known American hydrogen safety researchers is Michael Swain, an associate professor of mechanical engineering at the University of Miami at Coral Gables. For many years, Swain, whose involvement with hydrogen goes back to the early 1970s when he and a colleague converted a Toyota station wagon to hydrogen, has been looking systematically at safety issues. Summarizing past work at the 1998 DoE Hydrogen Program review in Alexandria, Virginia, Swain said that he and his colleagues had investigated hydrogen safety issues associated with automobiles, buildings, pipes, residential fuel lines and pipelines, hydrogen facilities of all types, ventilation systems, accidental combustion during hydrogen production, flame arresters, and hydrogen permeability.

At the DoE's 1996 Hydrogen Program review meeting in Miami, Swain reported that the energy leakage rate of hydrogen was less than that of natural gas or liquid petroleum gas in similar-sized leaks with the same line pressure, even if the line pressure was increased to produce the same energy flow rate as for natural gas or LPG. Swain compared the behavior and the potential danger of clouds of hydrogen with those of LPG via computer model analysis as well as experiments. He found that in most cases, LPG clouds were more dangerous than hydrogen clouds. This, Swain said, was due to different gas densities: the movement of a gas cloud in the air was dominated by the cloud's density, diffusion playing a secondary role. In examining various scenarios that could lead to accidents, Swain found that LPG, with its higher density, usually creates more danger than hydrogen does.

In 2001, Swain presented what may be one of the most convincing visual representations of hydrogen's greater safety yet—a video lasting three and a half minutes and a paper about a fuel leak simulation and ignition.[7] The video, shown at the National Hydrogen Association's annual meeting in Washington, D.C., showed the flames and fires, side by side, from one car with a hydrogen tank and one with a conventional gasoline tank. The hydrogen car leaked 3.4 pounds of hydrogen with an energy content of about 175,000 Btu, and the gasoline car leaked five pints of gasoline with about 70,000 Btu. Ignited about three seconds after the fuels began to leak, one set of pictures showed the hydrogen flame shooting straight up from the trunk area, while a big fire began to build underneath the gasoline car between the front and rear wheels.

After one minute, the hydrogen flow subsided, while the gasoline car began to be engulfed in flames that after another minute or so burned inside and outside the car. In the short paper describing the video, Swain said the maximum temperature measured on the hydrogen car was 117°F on the rear window glass, comparable perhaps to summer temperatures in the American Southwest, and a balmy 67°F on the rear tray, between the rear window and the rear seat. Summarized Swain succinctly, "The damage to the gasoline-powered vehicle was severe. The hydrogen-powered vehicle was undamaged."

In the late 1980s, the international hydrogen community began to recognize that hydrogen safety would have to be dealt with at the international level if hydrogen was ever going to become a global energy carrier. In addition, it had become clear to many that hydrogen safety is qualitatively different from the safety of other fuels and that international standards needed to be established. This led, in early 1990, to the establishment of a new hydrogen energy technical committee as part of the International Organization for Standardization, the Geneva-based international body that administers technical standards worldwide. The new group, known as Technical Committee 197, was organized largely through the efforts of a Swiss hydrogen supporter, Gustav Grob, who served as the first chairman. Switzerland relinquished the TC 197 secretariat in 1993, and Canada took over. In 2010 the secretariat, located in Sainte-Foy, near Quebec City, was coordinating work on more than a dozen technical reports and technical specifications covering subjects ranging from land vehicle fueling system interface, airport hydrogen fueling facility, hydrogen fuel specifications, compressed hydrogen surface, to hydrogen detection apparatus for stationary applications.

Postscript

There is no technical connection between the hydrogen bomb and hydrogen fuel. Hydrogen fuel represents chemical energy, combustion and burning, and, in fuel cells, electrochemical conversion processes; the hydrogen bomb works at the atomic level via principles of nuclear physics. But as Dan Brewer noted in his Ispra paper, the linkage provided by the word "is apparently enough to stir the imagination of the public and excite fear and suspicion of the fuel."

12

The Next Fifty Years

"As president, I will set a hard cap on all carbon emissions at a level that scientists say is necessary to curb global warming—an 80 percent reduction by 2050."

That was the key sentence in a 4,195-word campaign speech that presidential candidate Barack Obama gave before some seventy-five local environmentalists and activists on October 8, 2007, at the Portsmouth, New Hampshire, public library. In what was billed as a major policy address, Obama explained that if elected president, he would address environment and energy issues. He declared, "To ensure this isn't just talk, I will also commit to interim targets toward this goal in 2020, 2030 and 2040. These reductions will start immediately and we'll continue to follow the recommendations of top scientists to ensure that our targets are strong enough to meet the challenge we face."

Obama's plans had their ultimate roots in concepts in the U.N. Framework Convention on Climate Change going back to the late 1980s and 1990s. At a 1989 ministerial conference in Noordwijk, the Netherlands, for instance, the Dutch proposed that greenhouse gas concentrations should be stabilized at levels that would keep climate change within "tolerable limits" and that the Intergovernmental Panel on Climate Change (IPCC), which had been established the previous year, should report on possible options. The same year, a German investigating body, the so-called Enquete Commission, proposed an 80 percent reduction in fossil fuel use by 2050 to avoid warming of 1°C to 2°C. One undated IPPC graph, depicting the consistency of emissions trajectories with various atmospheric carbon dioxide (CO_2) concentrations, showed that emissions would have to drop to 450 parts per million to revert to 1990 levels; unchecked, they would climb to 750 parts per million. A note pasted into the graph stated, "Some scientists now say 350 ppm is necessary to avoid catastrophic climate change."[1]

These and other similar findings prompted the Obama administration's plans to launch a comprehensive climate change bill after it took office. The long-term case for hydrogen and fuel cells and other clean technologies to help clean up transportation was made forcefully in a series of reports by the National Research Council, which said in its initial 2008 main report that "the deepest cuts in oil use and CO_2 emissions after about 2040 would come from hydrogen." The committee that wrote the report listed as its top recommendation that hydrogen and fuel cell technology must be part of an array of technologies: "A portfolio of technologies including hydrogen fuel cell vehicles, improved efficiency of conventional vehicles, hybrids, and use of biofuels—in conjunction with required new policy drivers—has the potential to nearly eliminate gasoline use in light-duty vehicles by the middle of this century, while reducing fleet greenhouse gas emissions to less than 20% of current levels. This portfolio approach provides a hedge against potential shortfalls in any one technological approach and improves the probability that the U.S. can meet its energy and environmental goals." Another report, this one published in 2009, noted that fuel cell vehicles "could eliminate gasoline use by the fleet" of all vehicles traveling on American roads. And a final NRC report, published in June 2010, said the Freedom CAR public-private partnership to develop more fuel-efficient vehicles should continue to include fuel cells and other hydrogen technologies in its research and development portfolio. "Although it's important to work on near-term technologies, it's equally important for the partnership to perform the type of high-risk research in areas such as hydrogen that would not otherwise be taken on by the private sector, especially as the economy is still recovering," said committee chair Vernon P. Roan.[2] And a detailed analysis by C. E. Thomas, a widely respected analyst and expert (chapter 1), has shown that only hydrogen and fuel cells can achieve that 80 percent reduction of greenhouse gas emissions below 1990 levels (http://www.cleancaroptions.com/).

2010: Uncertain U.S. Prospects

But at the time of writing in fall 2010, before the November congressional elections, the picture and prospects for hydrogen and fuel cells in the United States looked somewhere between bleak and uncertain, with only a few bright rays in the gloom. The big climate change and energy bill had died that summer because of lacking votes in the Senate and opposition from both parties (An exhaustive thirteen-page article about the grisly "sausage-making" process of senatorial horse trading and jockeying that killed the bill ran in the October 11, 2010, issue of the *New Yorker* magazine). As explained by the U.S. Fuel Cell Council's Robert Rose in his presentation at the September 2010 "f-cell" symposium in Stuttgart, Germany, the post-2010 elections prospects looked doubtful: with big gains by the conservative, largely greenhouse-skeptical Republicans likely in 2010 and 2012, and with a president greatly diminished in the polls up for reelection in 2012, the chances for meaningful policy advances seemed dim.

As to the government's hydrogen and fuel cell programs, contrary to what one might expect, the picture did not look much better, Rose added. The Obama administration in 2009 tried to kill the federal program developing hydrogen as transportation fuel but was rebuffed by Congress, which restored most of the funding. In 2010, the program was moving along somewhat uncertainly with an administration request to cut R&D funding by 15 percent. But the guessing among some Capitol Hill watchers was that Congress would once again restore funding to perhaps $150 to $175 million for the proton exchange membrane (PEM) fuel cell program and maybe a little more than $300 million for all of the Department of Energy's (DoE's) hydrogen and fuel cell activities. The Obama administration's focus, exemplified by Secretary of Energy Steven Chu's views (chapter 1), remained focused on batteries for transportation, coupled to a strong research interest in biological and liquid fuels. But fuel cells remained in the mix, and there was a sense that policymakers were coming around to the view that fuel cells were needed in the long run to achieve policy goals, Rose believed.

As it happened, in summer 2011 President Obama himself announced his support for transportation fuel cells in his call for a big increase in automotive fuel cell efficiency, to 54.5 miles per gallon. A July 29 White House announcement said the administration was looking for "game changing" performance improvements, including "incentives for electric vehicles, plug-in hybrid electric vehicles, and fuel cell vehicles." The Environmental Protection Agency and the Department of Transportation's National Highway Safety Administration were tasked with setting up appropriate rules.

All this was in marked contrast to developments in Europe, Japan, and Korea. In Germany, carmakers—including Japan's Toyota—industrial gas providers, utilities, and the German national hydrogen and fuel cell organization NOW were forging ahead with government-backed plans outlined in fall 2009 to set up a national infrastructure network with perhaps 500 fueling stations and thousands, maybe tens of thousands, of hydrogen fuel cell cars by 2015. Europe's first large so-called Lighthouse hydrogen fuel cell project, "H2moves Scandinavia," was scheduled to get underway in 2011 with Oslo as its hub, including for starters seventeen fuel cell vehicles (Mercedes, Alfa-Romeo, Th!nk) and plans for hydrogen fueling stations ultimately stretching some 2,000 kilometers (1,250 miles) from Berlin, Germany, via Malmo, Sweden, and Copenhagen, Denmark, to Bergen on Norway's west coast. In Japan, more than a dozen major energy companies and carmakers got together in what an August 4 story in *Nikkei* described as a Hydrogen Supply Technology Association to begin laying the groundwork for a national hydrogen fueling station network; by 2015 Japan expects to have 100 stations operating, 1,000 stations by 2020, and 5,000 stations by 2030, reported Ikuo Kasahara, vice president for production engineering at Toyota Europe, at the September 2010 "f-cell" symposium in Stuttgart. South Korea had nine stations operating in 2010 and expects to have thirteen by 2011, with 100 vehicles already being driven by selected customers in 2010, said Joong-Hwan Jun, a scientist working on fuel cells at the country's RIST research institute, at the same conference. His commercialization road map projected pilot-scale production of 1,000 fourth-generation fuel cell vehicles annually in the 2012–2014 time frame, and 10,000 vehicles per year beginning in 2015. Even more astonishing were projections for China reported at "f-cell" by a German expert, Juergen Garche, of the Stuttgart-based Center for Solar Energy and Hydrogen Research. Drawing on information from Chinese colleagues, Garche's paper said China was expected to operate a fleet of some 110 million cars by 2020. Of these, he said later, some 4.4 million may be hydrogen fuel cell vehicles. "I think the Chinese are capable of that," Garche, who has visited China and has good contacts to Chinese experts, wrote in an e-mail.

1990s: High Hopes

These numbers and projections represent a quantum change from the late 1990s when the first tentative steps were taken toward hydrogen transportation and a hydrogen economy. A milestone marker was set in Iceland. Icy winds whipped snow and sleet through the streets of Reykjavik on February 17, 1999, whistling past the entrance of the luxurious Grand Hotel where some four dozen people—international business executives, energy technologists, environmentalists, Icelandic parliamentarians and politicians, and representatives of the local media—had assembled in a conference room for a press conference at which representatives of the Icelandic government, a newly formed Icelandic business consortium, and three multinational corporations—DaimlerChrysler, the Royal Dutch/Shell Group, and Norsk Hydro—announced a million-dollar joint venture, Iceland New Energy (INE), to investigate the potential of hydrogen as a clean substitute for fossil fuels in Iceland.

Hailing the project, Iceland's minister for environmental affairs, Gudmundur Bjarnason, urged the partners to "cooperate in a fruitful way." "Perhaps their work will even mark our everyday life in time to come," he said. The government, he declared, supported the project "in order to diversify the economy and lay the foundation for higher living standards in the future." The idea of trying out a prototype hydrogen economy on a large scale on an island such as Iceland had been around for decades. (In the 1970s and 1980s, Hawaii had been regarded as a good candidate.) A paper presented at the 1992 World Hydrogen Energy Conference by B. Arnason, T. Sigfusson, and V. Jonsson had laid out the specifics for the undertaking: Iceland's economically exploitable hydropower potential is about 30 terawatt-hours per year. (Its geothermal potential is much larger—about 200 terawatt-hours per year.) The large fishing fleet, which consumes about 230,000 tons of oil per year (about 12 percent of total consumption, and almost a third of all fossil-fuel imports), could serve as a demonstration project, with the oil being replaced by about 70,000 tons of hydrogen per year. And hydrogen energy could be a viable future export to Europe.

In an earlier memorandum of understanding, the parties agreed to cooperate in devising a road map to turn Iceland, with a population of about 313,000, into a laboratory for hydrogen energy technology and, within fifteen or twenty years, into a hydrogen economy. If things work out, eventually all of Iceland's cars and buses will run on either hydrogen or methanol, and the fishing fleet will be completely converted to fuel cell power, fueled by hydrogen derived from hydropower and geothermal energy. As it turned out, progress has not been quite as fast as had been hoped for in the early excitement. Reykjavik's mayor, Jon Gnarr, was given a fuel cell Ford Explorer in the summer of 2010 for an extended period, and a year earlier, an Icelandic Ford dealer had arranged to bring ten demonstration fuel cell Focus models to Iceland. In all, thirteen fuel cell vehicles are still operating in Iceland, but a couple of buses have stopped running, according to media reports. The *New York Times* said in a July 1, 2009, blog that the country's plans for hydrogen transportation had sunk "into the freezer," an assertion disputed by Maria Maack, INE's environmental manager. She acknowledged a slowdown, but that was due in part to a slowdown in deliveries of hydrogen internal combustion or fuel cell cars by participating manufacturers in Europe, Asia, and the United States, occasioned by the general global economic malaise. Deliveries of hydrogen cars that were supposed to arrive between 2006 and 2010 were going to be delayed until 2012 and 2015, she said. She insisted the goal was still to have a fleet of some 600 hydrogen cars on Icelandic roads by 2015, and if the transition comes a few years later, she asked, in an interview "Does that mean utter catastrophe if it varies by 5, 8, 10, or maybe 15 years?"[3] There is a chance that battery electrics may forge ahead in Iceland, given the country's big electricity-producing hydro and geothermal resources. In 2008, the country signed a memorandum of understanding with Mitsubishi to import small i-MiEV electric cars, but it is unclear where that stands now. And there is an Icelandic entrepreneur who wants to import electrics, starting with Revas made in India.

Since the beginning of the interest in hydrogen as an energy carrier, a vexing question had been how to get a hydrogen economy going. Endless debates about the proverbial chicken-and-egg problem—What should come first: economical hydrogen production or hydrogen-burning machinery?—have roiled the international hydrogen scene for decades. In the 1970s, the heyday of early hydrogen enthusiasm, one assumption was that hydrogen would somehow blend painlessly into the energy market. It was believed, for instance, that hydrogen could be mixed in ever-larger proportions into pipeline gas, eventually replacing natural gas altogether. However, it soon became clear that even if problems like pipeline embrittlement or availability of high-pressure compressors were solved, simple things such as metering the gas flow and energy content would present huge difficulties. (As it happened, one gas distributor, the Hawaii Gas Company, started doing just that in 2010 in a pilot project with General Motors.) Similarly, early hydrogen enthusiasts believed that hydrogen produced from water would find a ready market in ammonia production, for instance, as an entering wedge via the chemical industry. That did not happen, and in the 1990s, hydrogen produced via steam reforming from natural gas was still cheaper than hydrogen produced via water electrolysis.

1970s: A Long-Term Evolution

One view that developed in the 1970s among some hydrogen theoreticians was that the entire pattern of energy use, including the adoption of hydrogen, would evolve in three long stages. According to this scenario, in the first stage, coal in all its forms—mineral, liquid, gas—would gradually replace oil in twenty years or so, while light-water nuclear reactors would come into wider use and the first trials of breeder and high-temperature reactors would be conducted. Work on hydrogen would continue, but it would not make any significant inroads. In the second stage, a fifty-year period, a hydrogen economy "in the widest sense" would be phased in. Oil and natural gas would be largely eliminated as primary sources, replaced by solar and nuclear energy (high-temperature and breeder reactors). For cars, methanol rather than hydrogen would most likely be the most suitable fuel. Beginning some time around 2040, the third stage—the golden age of environmentally clean, planet-wide energy systems—would begin. Ocean mining would flourish, primary energy would come from solar sources of all types, and perhaps clean nuclear fusion power would come into the picture. Energy might be generated at a handful of very large sites far removed from population centers, the generated energy being transported to population centers as electricity (through long-distance, low-resistance cryogenic cables), hydrogen gas (in pipelines), or cryogenic liquid (in huge tankers).[4]

Some of these things have happened, if only in embryonic form, and others have not. Oil is still very much around, the oil industry seems bigger than ever before, and the conventional wisdom is that oil will be running the world for many decades to come. Coal is still the world's biggest energy source, but it is under attack. Nuclear fission, important in some countries such as France, is on the decline, and breeders and high-temperature reactors have not gotten off the ground.

Today, at the end of the twenty-first century's first decade, hydrogen is making slow progress, but widespread hydrogen use is still nowhere in sight. This is to be expected, at least according to the energy theorist and systems analyst Cesare Marchetti. The notion that phasing in hydrogen as a major energy carrier is an inherently long-term proposition, lasting from fifty to one hundred years, is grounded in Marchetti's 1970s investigations of the rate at which new energy systems—indeed, new major technologies of any type—replace older ones. Marchetti's curves and conclusions are still cited today, by, notably, Charlie Freese, General Motors's executive director in charge of fuel cell development.

Marchetti, long retired but probably still one of the most provocative long-range thinkers around, started from the "somehow iconoclastic hypotheses that the different primary energy sources are commodities competing for a market, like different brands of soap, or different processes to make steel, and that the rules of the game may, after all be the same," as he put it in a November 1974 lecture in Moscow.[5] He drew largely on the work of two American scientists, J. C. Fisher and R. H. Pry, who analyzed substitution rates of new technologies and attempted to forecast the rate of diffusion of new technologies for General Electric.

Using certain basic data and formulas, Marchetti found that he could predict market shares decades in advance. He produced some graphs that showed the market-penetration curves for various pairs of competing products or technological processes over periods ranging from sixty to eighty years—for example, the replacement of the open-hearth method of making steel by the Bessemer method, the substitution of sulfate turpentine for natural turpentine, and the replacement of water-based paint by oil-based paint. The astounding aspect of these graphs was that the equations describing market penetration worked well as a forecasting tool spanning decades. The curves derived from these formulas matched the historical data with "extraordinary precision," according to Marchetti.[6]

Marchetti decided to test the theory by trying to predict oil's percentage of the U.S. energy market, using data from four decades earlier. He told his Moscow audience that the results were astonishing:

I took the data for the US from 1930 to 1935 and tried to forecast oil coverage of the US market up to 1970. The predicted values, even for the saturation period, fit the statistical data better than one percent which, after all, is the minimum error that can be expected from this kind of statistics. This means that the contribution of oil to the US energy budget, e.g. in 1965, was completely predetermined 30 years before, with the only assumption that a new primary source of energy, e.g., nuclear, was not going to play a major role in the meantime. As the history of substitutions shows, however, the time a new source takes to make some inroads in the market is very long indeed—about a hundred years to become dominant, starting from scratch.

In 1976, at a microbial energy conversion seminar in Göttingen, Marchetti observed that the "takeover time" (the time it takes a new technology to go from 1 percent of the market to 99 percent) was about fifty years for the United States, an observation that presumably would be valid for hydrogen. "The extreme stability of the functions over very long periods of time, including wars, depressions, economic miracles, and a perceived acceleration of knowledge and 'progress'" do not detract from the validity of this phenomenon, Marchetti said. "My feeling is that all this is linked to learning processes at the societal and individual level which evolve very slowly and perhaps were the same 1000 years ago." "The takeover times," he observed, "are on the order of a century, so there is not much purpose in darting ahead. . . . Meaningful influences on the system can be obtained only through really long-term thinking and planning. . . . In fact, the spreading of a new technology always follows the rule of penetrating first small favorable eco-niches, acquiring force and momentum for the next step." The point, Marchetti told the audience, was to forget about heroic concerns like "the world energy market and world salvation" and to "concentrate on the special case, the special product, the favorable eco-niche." "A new technology," he said, "needs very special conditions to root."

Similar considerations, which still ring true today, led the authors of a massive 1976 Stanford Research Institute study, "The Hydrogen Economy: A Preliminary Technology Assessment," to argue that long-range changes affecting entire nations should not be left to private corporations, with their inherent short-term outlook. Corporations formally discount the value of future earnings because their philosophy is essentially piecemeal, thereby pushing the needed ultimate transformation further into the future; that, in turn, requires additional and essentially unnecessary changes at high cost, the authors said: "Decisions in both the private and public sector are generally made with a planning horizon of 5 to 10 years."

Because of the massive investments and long lead times required for phasing in a hydrogen economy, "corporations (and governments as well) tend to dismiss hydrogen in favor of concentrating on those activities that continue the viability of the existing order," the authors added. Here they were referring to the attention given during the 1970s to synthetic fuels made from oil shale and coal. Similar short-term strategies were still guiding some major corporations, and the U.S. government, in the late 1990s. The Chrysler Corporation, for example, before its 1998 merger with Daimler-Benz, signed an agreement with the Syntroleum Corporation of Tulsa to develop "designer fuels" manufactured from natural gas that would be inherently sulfur free and therefore more suitable for fuel cells than sulfur-containing fuels. The U.S. DoE also argued that it would make sense to extract hydrogen from gasoline and conducted R&D work in support of that objective. The basic rationale was that in the absence of a hydrogen refueling infrastructure and in view of the technical difficulties of storing hydrogen onboard vehicles, the existing gasoline distribution system should be used to provide fuel for fuel cell vehicles—hydrogen—with gasoline functioning as a hydrogen carrier.

But the big U.S. carmakers were losing their appetite for extracting hydrogen from gasoline. At the 1999 Detroit Auto Show, both DaimlerChrysler and Ford said that extracting hydrogen from gasoline was more difficult than had been anticipated a few years earlier and that it was an immature technology that would take a long time to get to market.

As the 1990s drew to a close, there were signs that the admonitions of hydrogen supporters to look seriously at hydrogen as a clean energy option for the world were beginning to be heeded. Probably the best example was Japan's much-commented-on WE-NET (World Energy Network) project, which, more than any other national scheme, recognized the need for long-term planning and investment in hydrogen energy technology. The $2 billion WE-NET project—part of the Global Warming Prevention Plan adopted by Japan's cabinet in October 1990 after the International Panel on Climate Change targeted CO_2 emission reductions at the 1990 level by the year 2000—was announced in 1993 and was to run for twenty-eight years, through 2020. There were nine major subtasks during the initial five-year Phase I, which ran from 1993 to spring 1999, including water electrolysis for hydrogen production, conversion of gaseous hydrogen into other forms for transport, and fuel cell development. Forty entities, including universities, think tanks, and corporations, were taking part in Phase I. Phase II, which began in April 1999 and ran through Japan's fiscal year 2003, was essentially a continuation of Phase I. As it turned out, WE-NET ended after less than a decade, replaced by a wide array of other programs with similar objectives run by other entities; its final annual report was produced in 2002, according to the WE-NET Web site, which was still up in 2010 (http://www.enaa.or.jp/WE-NET/).

1997: The President's Advisers: Do More in Hydrogen

In November 1997, a month before the Kyoto Global Climate Change Conference, a top-level advisory group to President Clinton told the president that the United States should do more in the areas of renewable energy and hydrogen technology. The twenty-one-member energy panel of the President's Committee of Advisors on Science and Technology (PCAST), the highest-level private sector science and technology advisory group, had been set up to review the nation's energy priorities, which had not come under scrutiny in many years. It recommended doubling spending of the DoE's hydrogen program as part of a general shifting of the department's priorities toward greater emphasis on renewables and energy efficiency.

In his cover letter to President Clinton, John Gibbons, director of the White House Office of Science and Technology and PCAST's cochairman, said the country needed to improve its R&D effort, especially "in relation to the challenge of responding responsibly and cost-effectively to the risk of global climate change from society's greenhouse gas emissions, in particular, carbon dioxide from combustion of fossil fuels."

In January 1998, the DoE published a five-year strategic plan for its hydrogen program outlining what the program's managers expected to be the program's goals and hopes over the next twenty years. "Dependence on foreign energy sources is expensive," according to the introduction. "We suffer trade deficits and use our military to protect our energy supply abroad. Environmentally, the Nation is being forced to react to both the need for cleaner urban air and to the potential effects of global climate change. . . . The solution is a clean, sustainable, domestic energy supply. Hydrogen can be one of the answers." Among its goals, the plan listed improving the efficiency and lowering the costs of fossil-fuel-based and biomass-based hydrogen-production processes to $6 to $8 per million Btu, developing renewable-based, emission-free hydrogen production technologies with a target cost of $10 to $15 per million Btu, developing fuel cell and reversible fuel cell technologies, and supporting industry in the development and demonstration of hydrogen systems in the utility and transportation sectors.

In Europe, two think tanks, the London-based World Energy Council (WEC) and the International Institute for Applied Systems Analysis (IIASA) in Laxenburg, Austria, collaborated in an effort to see how the world's energy future might unfold in the twenty-first century. Their joint 1996 study, "Global Energy Perspectives to 2050 and Beyond," concluded that the choices about the types of energy that will be used in future decades will be made in the first two decades of this century. As incomes increase around the world and as the world's population roughly doubles by midcentury, "people will want higher levels of more efficient, cleaner and less obtrusive energy services," said the authors, Arnulf Grübler and Nebojša Nakićenović of IIASA and Michael Jefferson of WEC. Global demand for energy services will grow by as much as an order of magnitude by 2050, but primary energy demands will grow less because of improvements in energy intensities—more primary energy input into a given product or service, replacing physical materials. The report stressed "decarbonization," that is, removal of carbon from fossil fuels by the twin technologies of carbon capture and carbon sequestration, the safe storage of CO_2. Both are being tested by laboratories and research institutions around the globe.

What is left is, of course, hydrogen. Decarbonization will be the industrial end-game strategy of a trend first detected by Cesare Marchetti in the 1970s when he described a gradual shift, over centuries, from hydrocarbon fuels with high carbon and low hydrogen content (wood, peat, coal) to fuels with increasingly less carbon and more hydrogen (oil, natural gas), culminating, seemingly inevitably, in pure hydrogen as the principal energy carrier of an advanced industrial society. A diagram in an another Nakićenović essay, "Freeing Energy from Carbon," in the summer 1996 issue of *Daedalus*, showed how the carbon intensity of the world's energy consumption, expressed in tons of carbon per ton of oil equivalent energy, dropped from just below 1.1 in 1860 to 0.7 in the early 1990s—a decline of about 0.3 percent per year. "The ratio has decreased because high-carbon fuels, such as wood and coal, have been continuously replaced by those with lower carbon content, such as gas, and also in recent decades by nuclear energy from uranium and hydropower, which contain no carbon," Nakićenović said. In a 2010 update, Peter Kolp, a Nakićenović colleague at IIASA, provided charts for historical trends from 1971 to 2007. They showed that the carbon intensity of the gross domestic product, that is, economic activity, steadily declined from about 0.24 percent to about 0.14 percent (in terms of purchasing power parity); energy intensity, or units of energy per unit of gross domestic product, declined from about 0.32 percent to just below 0.2 percent in the same period.

In "On the Patterns of Diffusion of Innovation," also published in the summer 1996 issue of *Daedalus*, Arnulf Grübler asserted that there are basically two innovation strategies. One focuses on incremental changes such as environmental add-on or "end-of-pipe" technologies. These can provide a quick fix, but they "tend to reinforce the dominant trajectory, blocking more systemic and radical changes." The other strategy opts for "more radical departures from existing technologies and practices." These, "such as the development of fuel cells and hydrogen for energy," are "more effective in the long run"; however, they "require much more time to implement because of the multiplicity of forward and backward linkages between technologies, infrastructures, and forms of organization for their production and use." According to Grübler, this interdependence between "individual artifacts and long-lived infrastructure" is at the root of the problem we face:

Within two or three decades, the United States could in principle change its entire fleet to zero-emission vehicles. In fact, 99 percent of vehicles now on the road will be scrapped in this interval. Yet, this interval is too short for the diffusion of the required associated energy supply, transport, and delivery infrastructures, which will inevitably distend the rate of diffusion of end-use devices. Thus, key technologies that we can already envision to raise the quality of the environment probably must await the second half of the twenty-first century to become widespread and influential.

Historically, technology clusters (parallel development of railroads and telegraph, road networks, oil pipelines) have been instrumental in raising productivity and also in alleviating many adverse environmental effects. The emergence of a new cluster could hold the promise of an environmentally more compatible technology trajectory. But it will take time.

Exactly when will that cleaner, more efficient energy be provided? Theorists looking at planetary energy issues believe it will take a long time.

"The question of what kind of companies will supply energy services, and how, is wide open," according to the 1996 WEC-IIASA study. Until about 2020 there will not be much of a change owing to the long lifetimes of power plants, refineries, and other energy facilities and investments. But during those first twenty years of the twenty-first century, the choices that will determine the physical characteristics of the world's energy economy will likely be made, and those choices will largely determine the course of the following eighty years.

The WEC-IIASA study tried to imagine a wide range of possible pathways to the future. It assumed three basic cases, subdivided into six scenarios, each an "alternative image of how the future could unfold." Assumptions ranged from "a tremendous expansion of coal production to strict limits, from a phase out of nuclear energy to a substantial increase, from carbon emissions that are only one-third of today's level to increases by a factor of more than three." All six scenarios foresee a shift toward electricity and to higher-quality fuels, such as natural gas, oil products, methanol, and "ultimately, hydrogen":

No matter what the eventual dominant fuel in a scenario, there is a shift away from noncommercial and mostly unsustainable uses of biomass, and direct uses of coal virtually disappear. Fossil sources continue to provide most of the world's energy well into the next century but to a varying extent across the scenario. Sustainable uses of renewables come to hold a prominent place in all scenarios.

In another study, "Decarbonizing the Global Energy System,"[7] Grübler and Nakićenović concluded that decarbonization proceeds very slowly, at an average rate of 0.3 percent a year. If the trend holds true, "we might in fact be only half-way through the fossil fuel age that would draw to a close only late in the 22nd century." Globally, the authors said, the energy system is moving in the right direction. But "it will simply not suffice to rely on 'autonomous' structural change toward carbon-freer energy systems. . . . [The energy systems] are dwarfed by historical and anticipated growth rates in energy use and resulting carbon emissions. Substantial acceleration of decarbonization would thus entail both ambitious technological and policy changes."

Cesare Marchetti believed that hydrogen and the emergence of a hydrogen economy have a deeper meaning that goes beyond clean fuel and clean environments. Back in the 1970s, he speculated on the idea that the physical changes associated with making and distributing hydrogen energy would fundamentally affect global politics. Marchetti outlined his ideas at the 1976 Miami Beach hydrogen conference in a forty-minute address titled, "From the Primeval Soup to World Government—An Essay on Comparative Evolution." He characterized his startling politico-economic predictions as logical outcomes of the development of a hydrogen economy and as analogous to the physical processes that have shaped the Earth since its beginning. He envisioned large-scale production of hydrogen and a move away from fossil fuels as a planetary liberating process almost comparable to the move away from the primeval soup as energy and life carrier millions of years ago. That move was made possible by the release of tremendous amounts of oxygen through a series of complicated chemical and physical processes. Oxygen formed the ozone layer and, by filtering out lethal ultraviolet radiation, allowed living organisms to come out of the water and conquer the land. "Life had grown to take control of the environment on a global scale," Marchetti asserted. "Now," he continued, "if humanity starts producing hydrogen from water using 'new' sources of free energy—fission, fusion, or perhaps directly tapping the old sun—the moves away from fossil fuels which in my analogy are the equivalent of the primeval soup, what kind of new control of the environment can occur as by-products of this operation?"

In essence, Marchetti foresaw a conflict between the "vertically integrated" power system of nation-states and the "horizontal" power system, covering many states, typical of the multinational corporation. Conflict is inevitable, he believed, but it would not necessarily lead to the destruction of either the states or the corporations. "A state can be seen as a vertically integrated power system filling a geographical area," Marchetti argued. "A multinational is a horizontal power system organizing a thin layer of human activities without precise geographical boundaries. The horizontal power (the multinational) generates 'confusion' and loss of control by the vertical one (the state), and, with the layers thickening, will inevitably lock with it. The ostensible muck-raking against multinationals is a clear symptom of that. But what will be the outcome?"

Marchetti cited as a historical analogy the relations between the single "multinational" of medieval times, the Catholic church, and the states of those times. "A fight between geographically bound political powers and a pervasive horizontal power went on in the Middle Ages," he said. The church was "interfering and competing in many ways with the geographically fragmented political power of the time." Both systems were necessary: "As the two power systems cannot be interchanged, nor either one eliminated, a compromise finally had to be worked out. The political power handed its thin top layer to a supernational power structure, producing a kind of political multinational: the Holy Roman Empire. The maneuvering space of the emperor was politically narrow, but territorially broad and at the proper hierarchical level of abstraction to deal with the Pope on an even basis." "My educated analogy," declared Marchetti, "says that the outcome will be a world government, or more flexibly defined, a world authority to make a dialogue possible." "Energy," he added, "is the largest single business in the world. . . . Energy multinationals will be the strongest forces in the struggle with political power, and their field of activity a sensitive one. Very large energy centers and energy generation as a world operation are direct consequences of the technological process of water-splitting, and they may modify the political atmosphere, just as oxygen did for the earth atmosphere, leading finally to a world government."

New Post-2000 Voices: Rifkin, Burns, Hofmeister

A quarter-century later, some of Marchetti's notions resonated in Jeremy Rifkin's 2002 book, *The Hydrogen Economy.*[8] Rifkin, an economist and social critic, runs the Foundation on Economic Trends in Bethesda, Maryland, and he thinks in cosmic terms. Rifkin, who has been advising, among others, German chancellor Angela Merkel, Italy's former prime minister and former head of the European Commission, Romano Prodi, believes we are on the cusp of a Third Industrial Revolution, spearheaded by renewable energy and hydrogen technology in all its variations, including fuel cells. He has pronounced, for example, "The really great economic revolutions in history occur when new communications technologies fuse with new energy regimes to create a wholly new economic paradigm," and, "Hydrogen, because of its universality, offers the prospect that we might be able, at long last, to democratize energy and empower every human being on Earth." He paints a somber picture of a world running out of oil and teetering on economic collapse—a notion scoffed at by some critics when the book came out, but gaining credibility five years later with the onset of the global financial and economic crisis in 2007 whose aftershocks are still felt today, in 2010. With oil dictating the activities of commercial and governmental institutions everywhere, Rifkin believes the solution lies in a worldwide hydrogen energy web that links energy producers and consumers with fluctuations in supply and demand managed by a computerized grid that tracks demand changes everywhere, moving excess capacity and output to where it is needed, enabled by the interchangeability of hydrogen and electricity—the hydricity concept first enunciated by the late Geoffrey Ballard (chapter 1).

A similar theme, "the need to transform the automobile's DNA," was struck in a 2010 book by two automobile industry experts and an architect and city planner. *Reinventing the Automobile* was written by Lawrence D. Burns, who, before his 2009 retirement as General Motors vice president of research and development, had been propounding for years the need for automotive DNA transformation; Christopher E. Borroni-Bird, who managed GM's PUMA (Personal Urban Mobility and Accessibility) project of a gyro-balanced two-wheel battery-powered city vehicle (a rickshaw-like Segway for two); and the late William J. Mitchell, who directed MIT's Smart Cities project, a research effort that pursues sustainability, livability, and social equity through technological and design innovation.[9] It described the vision of transforming our mechanical conveyance, the car, powered by fossil fuel combustion into a small, smart, electronic appliance powered by a fuel cell or grid electricity and linked to all other vehicles for maximum traffic efficiency and safety by a computerized Internet information system. While battery vehicles offer "excellent" performance in small urban vehicles with zero emissions, their range is limited and "refueling"—recharging the batteries—takes hours. "Only the hydrogen fuel cell option promises to combine the range and refueling-time convenience of conventional family-sized vehicles with the energy and environmental benefits of pure battery-powered vehicles," they wrote. "Moreover, the hydrogen infrastructure complements the electric grid from an energy-density perspective because domestically supplied reformed natural gas and biomass are excellent sources of hydrogen, and hydrogen, generated from electrolysis of water, is an excellent way to store electricity produced from renewable sources like the wind and the sun." Among other things, they noted, several government and industry studies have concluded that a hydrogen infrastructure "is economically viable and technically feasible. In fact, for less money than was spent to build the Alaskan pipeline, conveniently located hydrogen stations could be deployed in the 100 largest U.S. cities and every 25 miles on all interstates, putting hydrogen conveniently in reach for 70 percent of the U.S. population."

Support for these new technologies also comes from a third, perhaps unexpected, corner: the retired CEO of a major oil company. John Hofmeister, until his retirement in 2008 as president of Shell Oil Company and U.S. Country Chair, says in his 2010 book, *Why We Hate the Oil Companies,* "We need to look at radical shifts away from our dependence on the internal combustion engine and pulverized coal-fired electricity plants.[10] Hydrogen fuel cells and niche battery applications work for transportation," while nuclear power and gasified coal "with carbon capture and sequestration work for electricity, along with wind and future solar, augmented by natural gas." Hofmeister, a member of DoE's Hydrogen and Fuel Cell Technical Advisory Committee and founder of the nonprofit Citizens for Affordable Energy, wrote that "the abundance of Earth's energy resources, from fossil fuels to non-carbon fuels, renewable and manufactured, is unquestionably greater than we commonly know." He added, "We will never lack for energy in the future because of scarcity of resources. We will, however, face serious shortages of energy in the future unless we start making better, often harder, choices of what will supply our future needs. Such supplies will also need to work in an environmentally challenged world." If the United States were serious about energy reform, which he clearly thinks we are not, "wouldn't we also be building the manufacturing infrastructure to construct the entire new energy system?" Meanwhile, the country watches "Japan and Germany build a totally new hydrogen infrastructure to support introduction of the fuel cell technology they are developing to replace the internal combustion engine." Hofmeister, who distinguishes between "political time" and "energy time," believes that beyond hydrogen and fuel cells, the United States needs an independent federal energy resources board, body patterned after the Federal Reserve Board, to regulate energy supply, technology choices, environmental protection, and infrastructure choices, including a supply system that "may well require a plan that stretches out in 'energy time' from now to 2060. Such a five-decade plan is impossible to conceive in 'political time' because it would span at least 25 Congresses and as many as 6 to 12 presidents." Warned Hofmeister, "The time "to begin deliberations on the creation of a new federal independent regulatory agency is now—sooner, not later—before the United States is in dire crisis, before endemic shortages and unprecedented price spikes have further poisoned the national dialogue, before more so-called oil wars or geopolitical rivalries arbitrarily reduce energy supplies and more seriously divide the nation into energy 'haves' and 'have-nots.'" Hofmeister's fifty-year road map vision includes the build-out of new manufacturing infrastructure and a new car fleet of battery electrics and hydrogen fuel cells gaining momentum in the medium term, including a supply of hydrogen fuel from a variety of sources and including a shift to lower-carbon ones.

There has been a great deal of progress since 1990 in terms of greater environmental awareness and the cleaning up of large parts of our environment. As noted throughout this book, much has also been achieved toward the development of clean, environmentally progressive alternative technologies and practices. And in the past fifteen years or so, we have witnessed an acceleration of the race to bring hydrogen-based zero-emission or near-zero-emission technologies such as fuel cells to

Notes

Chapter 1

1. R. Buckminster Fuller, *Utopia or Oblivion: The Prospects for Humanity* (New York: Overlook, 1969).

2. United Press International, summarizing a Saudi television broadcast of December 21, 1976, after the OPEC conference at Doha, Qatar.

3. From a transcript of Clinton's remarks at the White House Conference on Climate Change, October 6, 1997. In fact Toyota was not working with Ballard Power Systems but was developing its own fuel cells.

4. Federal Energy Research and Development for the Challenges of the Twenty-First Century, *Report of the Energy Research and Development Panel*, November 5, 1997.

5. "Moving Slowly toward Energy Free of Carbon," *New York Times*, October 31, 1999.

6. Chu spoke at a May 7, 2009, press briefing discussing the Department of Energy's (DoE) 2010 budget request. He requested a cut of about $130 million, which would have killed the transportation part of DoE's hydrogen and fuel cell program if it had been enacted. The request created a firestorm of protest from groups as diverse as the Alliance of Automobile Manufacturers, the American Lung Association, the Electric Drive Association, the Union of Concerned Scientists, the National Research Council, the National Hydrogen Association, and the U.S. Fuel Cell Council. Byron McCormick, the retired head of General Motors's fuel cell program, resigned in protest from DoE's hydrogen and fuel cell advisory committee. Senator Byron Dorgan (D, North Dakota), a strong hydrogen supporter, told Chu at an appropriations subcommittee hearing he chaired that he was "stunned" by the cuts, that they were a "significant mistake" and "not a smart thing to do." Congress agreed and restored almost all of the cuts later that year. Chu made another attempt to cut funding the following year, only to be rebuffed again. In late 2010, there were some signs that Chu and DoE's leadership were coming around to the idea that much, perhaps most, of the rest of the world was embracing hydrogen for transportation and other uses and that the United States was in danger of being left behind, but these hopes were dashed in 2011 when Chu again proposed cuts in key hydrogen and fuel

cell programs. However, in July 2011 it looked as if Chu's boss, President Obama, was having second thoughts and was coming around to supporting fuel cells and other electric vehicle technologies by ordering up incentives for their development (see chapter 12).

7. "The Role of Battery Electric Vehicles, Plug-In Hybrids and Fuel Cell Electric Vehicles" (New York: McKinsey, November 2010).

8. The 1998 *Oxford Illustrated Dictionary* defines *pollute* as to "contaminate or defile [the environment]." Hydrogen, when combined (burned, oxidized) with the air's oxygen, produces only water plus minuscule amounts of oxides of nitrogen, inevitable by-products of any atmospheric burning process. Water does not pollute.

9. Others dispute that there is much of a problem. Among academics, one of the most prominent and respected skeptics is meteorologist Richard Lindzen of the Massachusetts Institute of Technology, who says flatly, "We don't have any evidence that this is a serious problem" (in Global Warming, *New York Times* supplement on Kyoto Climate Change conference, December 1, 1997). Lindzen, described in the *Times* as a "champion to political conservatives and industrial interests who minimize the threat" but also as "a force of intellectual honesty in a highly politicized debate," told the *Times* that he prizes the environment but that global warming and other issues have prompted environmental groups to go "off the deep end" and produce "a drum roll that gets rid of perspective." Lindzen's critics, said the *Times*, fault him for professing unwarranted sureness in a field of research rife with uncertainty, and many say he is simply wrong. Lindzen has consistently stuck to his views. In 2009, he told the March 8–10 International Conference on Climate Change in New York City, "There is no substantive basis for predictions of sizeable global warming due to observed increases in minor greenhouse gases such as carbon dioxide, methane and chlorofluorocarbons."

10. Hans Deuel, Paul Guthrie, William Moody, Leland Deck, Stephen Lange, Farhan Hameed, Jeremy Castle, and Linda Mearns, "Potential Impacts of Climate Change on Air Quality and Human Health" (paper presented at the 92nd Annual Meeting of Air and Waste Management Association, St. Louis, 1999).

11. A quad is equal to 8 billion gallons of gasoline—enough to run about 14.5 million cars for a year.

12. About 78 percent of the air we breathe is inert nitrogen; about 21 percent is oxygen. Air also contains trace gases, including CO_2 (0.03 percent).

13. For a more detailed explanation of how fuel cells work, see chapter 7.

14. Clean Car Options, http://www.cleancaroptions.com/html/vehicle_fuel _economy.html. The fuel cell version of the Highlander was certified by DoE national laboratory engineers at 68.3 miles per kilogram of hydrogen, which is equivalent to 69.1 miles per gallon of gasoline equivalent—3.1 times the fuel economy of the conventional Highlander SUV.

15. *Encyclopedia of Chemistry*, 3rd ed. (New York: Van Nostrand Reinhold, 1973), 544.

16. For decades, BMW has been the principal supporter and advocate of this approach for onboard storage in cars.

17. Some earlier advocates have stated that over distances of more than 600 miles, hydrogen pipelines transport energy cheaper than high-voltage (500 kilovolts) overhead AC transmission lines. See O'M. Bockris, *Energy Options: Real Economics and the Solar-Hydrogen System* (Sydney, Australia and New Zealand Book Company, 1980), 216.

18. However, there is a serious drawback associated with methanol, also known as wood alcohol or methyl alcohol, and used in industry as a solvent, an antifreeze, a denaturant for ethyl alcohol, and a raw material in the synthesis of formaldehyde and other chemicals: it is acutely toxic.

19. Personal e-mail communication, June 28, 2011.

20. E-mail communication, September 9, 2009.

21. Al Gore, *Earth in the Balance: Ecology and the Human Spirit* (Boston: Houghton Mifflin, 1992)

22. Ross Gelbspan, *The Heat Is On* (Reading, MA: Addison-Wesley, 1997).

Chapter 2

1. Julius Ruska, in *Das Buch der grossen Chemiker*, ed. G. Bugge (Weinheim: Chemie Verlag, 1974), 2.

2. Steven Weinberg, *The First Three Minutes* (New York: Bantam Books, 1977), 114.

3. E. Pilgrim, *Entdeckung der Elemente* (Mundus, 1950), 144.

4. Nitro-aerial because they occur also during combustion of saltpeter.

5. Richard Koch, in *Das Buch der grossen Chemiker*, 194.

6. Pilgrim, *Entdeckung der Elemente*, 155.

7. Georg Lockemann, in *Das Buch der grossen Chemiker*, 256.

8. Georges Cuvier, in *Great Chemists*, ed. E. Farber (Hoboken, NJ: Wiley-Interscience, 1961).

9. *Montgolfières* were hot-air balloons named after their inventors, the Montgolfier brothers. They were kept aloft by hot air generated by an open fire in the craft's gondola, not by a buoyant gas such as hydrogen. They were the first balloons ever to rise into the air. In their first flight, launched June 5, 1783, in Annonay, near Lyons, the Montgolfiers' craft stayed aloft for ten minutes and covered more than a mile. That same year, a *montgolfière* sailed over Paris in the first manned free balloon flight.

10. J. Pottier and C. Bailleux, "Hydrogen: A Gas of the Past, Present and Future," in *Proceedings of Hydrogen Energy Progress VI*, vol. 1, ed. T. N. Veziroglu et al. (New York: Pergamon, 1986).

Chapter 3

1. W. Cecil, "On the Application of Hydrogen Gas to Produce Moving Power in Machinery," *Transactions of the Cambridge Philosophical Society* 1 (1820, 217–239).

2. My source for Jules Verne's *The Mysterious Island* is the 1965 Airmont Publishing Company edition.

3. Max Pemberton, *The Iron Pirate* (London: Cassell and Company, 1893).

4. J. B. S. Haldane, *Daedalus or Science and the Future* (New York: Dutton, 1925).

5. The latter was a reference to the first fuel cell, constructed in 1839 by the English philosopher-physicist-lawyer William Grove, which produced water from oxygen and hydrogen but almost no electricity. Development of modern fuel cells that also produce electricity did not start until the 1950s.

6. D. Cosci, Impiego dell' idrogeno nei motori a combustione interna. *Rivista Aeronautica* 13, no. 2 (1937): 253–286.

7. I. I. Sikorsky, Science and The Future of Aviation, April 1938.

8. The Fischer-Tropsch process synthesizes various hydrocarbons from carbon monoxide and hydrogen. The Bergius process, which breaks down coal into a synthetic crude oil with the help of hydrogen and a catalyst, enabled Nazi Germany to produce large amounts of synthetic aviation gasoline.

9. There was a happy footnote, however. In 1978, Erren showed up at the World Hydrogen Energy Conference in Zurich. When he was introduced at the final session, the applause was loud and long, a fitting tribute for a man who had been forty years ahead of everybody else.

10. One exception was R. O. King, who continued his research on hydrogen combustion at the University of Toronto in the late 1940s and the 1950s.

11. Eduard W. Justi and August W. Winsel, *Kalte Verbrennung, fuel cells. Cold combustion, fuel cells.* German or English. (Wiesbaden: Franz Steiner Verlag G.M.B.H, 1962).

12. Eduard Justi, *Leitungsmechanismus und Energieumwandlung in Festkoerpern* (Goettingen:Vandenhoeck & Ruprecht, 1965).

13. J. O'M. Bockris, *Energy: The Solar-Hydrogen Alternative* (New York: Halstead Press/John Wiley & Sons, 1975).

14. Marchetti's remarks, "Proteus vs. Procustes," exist only as a sixteen-page text, plus two pages of tables and endnotes, in the author's possession. Marchetti, who is retired and lives near Florence, Italy, wrote in July 4 and 12, 2011 e-mails that he suspects "this was one of my free-wheel speeches without written support." He doesn't recall the exact date but remembers he was invited by Thomas Gold, an unorthodox and at times controversial astronomer at Cornell who, among other things, had developed a theory that both abiogenic methane and hydrogen are generated deep in the earth crust (*The Hydrogen Letter*, January 1987).

15. Marchetti's interest in hydrogen started in a roundabout fashion in the late 1950s, when he was working for the Italian national oil company Ente Nazionale ldrocarburi, trying to find ways to make use of the off-peak power of nuclear reactors. "The idea then was to somehow use electricity to make adenosine triphosphate (ATP) which is the energy carrier of biological systems," he recalled in an early interview. Marchetti tried to find ways to make synthetic ATP to act as an intermediary between an electrical energy source, like a reactor, and a biological system. Eventually Marchetti found "this didn't work because ATP never leaves the individual cell, therefore you cannot put it into a cell." In the 1960s, he recalled that in his high school days, he had read that there were certain bacteria able to metabolize hydrogen and oxygen and grow on that: "So I thought this material would be the right interface between food chain and electricity because you can make hydrogen and oxygen electrolytically." A bibliographical search of the literature under H—for hydrogen—turned up all kinds of chemical processes "that could do almost everything—reduce materials in chemistry, reduce minerals to metals, run engines and things like that." Looking at the problem more generally, "I decided from a systems point of view electrolysis was not very good for a number of reasons. So, going to the root of the problem, that of transferring free energy from a heat source to a chemical system, I decided that the best way was to go to a chemical system via a multi-step process." That was in about 1969, and the way he tells it with a smile, "I told De Beni to invent such a system, and he did invent one."

16. Linden was one of the first energy experts to use the term *hydrogen economy*. See Henry R. Linden, The Hydrogen Economy, *Journal of Fuel Heat Technology* 18 (1971): 17.

17. I'm not sure, but I think it was in a conversation with me, circa 1979.

18. Lawrence W. Jones, "Liquid Hydrogen as a Fuel for the Future" *Science* 22, no. 4007 (October 1971): 367–370.

19. Lawrence W. Jones, "The Hydrogen Fuel Economy: An Early Retrospective" *Journal of Environmental Planning and Pollution Control* (1973).

20. I was working at the time in Milan, Italy, as bureau chief for *McGraw-Hill World News*, the company's in-house news service for most McGraw-Hill technical and business magazines, including *Business Week*. I had suggested the piece after interviewing Marchetti at the Euratom Research Center in Ispra, some 40 miles up the road from Milan in the foothills of the Italian Alps. That taped interview in which Marchetti laid out the basic concept of a hydrogen economy marked the beginning of my interest in hydrogen.

21. In a 1976 interview, Escher recalled: "I became very much interested in hydrogen because it was paying my salary in 1958 when I began work on hydrogen-oxygen rockets at NASA Lewis [Research Center]. My job was to help run tests and lay out thrust chambers that worked on hydrogen and oxygen which was considered a far-out rocket fuel." Escher's "transition to real life," as he called it, came in the late 1960s, when he went to work for Rocketdyne, a maker of rocket equipment in California, and was assigned to sell rocket technology to the outside world. He recalled: "I made a few trips to the public utilities

and other such people with potential hydrogen interests. It was interesting, and immediately it churned up into a big activity, and a year later, in 1970 I left and became an independent consultant [on hydrogen and hydrogen-related alternative energy systems]." At the end of the 1990s, after several career moves, Escher joined the Science Applications International Corporation in Huntsville, Alabama.

22. Bockris, Escher, Marchetti, and Weil have already been introduced in this book. Tokio Ohta taught in the Department of Electrical Engineering at Yokohama National University in Japan; Van Vorst lectured at the School of Engineering and Applied Science at the University of California at Los Angeles; Martinez worked for Venezuela's national oil organization in Caracas; Seifritz was a lecturer at the Swiss Federal Institute for Reactor Research; Abdel-Aal was with the College of Petroleum and Minerals in Dharan, Saudi Arabia.

23. The DoE, established in 1977, combined several other agencies, including the Energy Research and Development Administration, set up in 1974.

24. Project Sunshine's other main research areas were solar energy, geothermal energy, coal gasification, and liquefaction.

25. Senators Spark Matsunaga (D, Hawaii) and Daniel Evans (R, Washington) were the Senate's most prominent hydrogen advocates at the time.

26. *The Hydrogen Letter*, May 1986.

27. A real foul-up occurred in 1986 when $1 million earmarked for the hydrogen program at Brookhaven National Laboratory vanished in the congressional budgeting process. The word from Capitol Hill was that this was due to an editing error, not because anybody wanted to scuttle the hydrogen program. "It fell between the cracks," said a Capitol Hill source at the time. Ultimately the missing million was restored. After 1987, it took ten years to restore the DoE's hydrogen program to a semblance of reasonable funding, with $15 million appropriated in both 1996 and 1997—still small sums compared to those for other renewables such as photovoltaics, which received close to $60 million in fiscal 1997. (In fairness, I should mention that hydrogen-related areas such as fuel cells got in the neighborhood of $70 million in fiscal year 1998—about one-third for transportation and the rest for stationary fuel cells.).

28. In the past, others have estimated higher real costs, long before oil prices hit the astronomical figures of $147 a barrel in mid-2008. In a December 1997 article in the *Atlantic Monthly*, Paul Ehrlich and four coauthors say the price of gasoline carries "a social cost of at least $4.00 a gallon but is sold to Americans for $1.20." Senator Richard Lugar (R, Indiana), chairman of the Senate Agriculture, Nutrition and Forestry Committee, estimated the real cost of Middle Eastern crude oil to be at least four times the going price. "The effects of world dependence on Middle Eastern oil means that while the quoted market price per barrel is about $20, the cost associated with keeping shipping lanes open, rogue states in check and terrorists at bay, may more than quadruple the price per barrel," Lugar said in his opening statement at a November 13, 1997, hearing of the Committee on Energy Security and Global Warming.

Chapter 4

1. Zweig cited data published in the mid-1990s by Jane Hall of Fullerton State College and David Abbey of the Loma Linda Medical Center School of Public Health showing that the societal cost of these illnesses was about $14 billion per year in the Los Angeles Basin alone. Zweig died in 2002 at the age of seventy-seven.

2. *Hydrogen & Fuel Cell Letter*, October 2009.

3. *Hydrogen & Fuel Cell Letter*, June 2009.

4. Wikipedia's entry, "Hydrogen Production," http://en.wikipedia.org/wiki/Hydrogen_production, gives a succinct overview of most hydrogen production methods in use or being researched, including a number of links and references.

5. Institute of Gas Technology, Survey of Hydrogen Production and Utilization Methods, 1975.

6. The energy content of fuel gases is typically expressed as higher (gross) heating value (HHV) or lower (net) heating value (LHV). Both measure the heat—number of British thermal units (Btu)—produced per standard cubic foot (scf) of the gas in the complete combustion of the gas, at constant pressure, and with the product of the combustion measured at 600°F and all water formed by that combustion condensed into liquid again. The lower heating value uses the same data, but the water does not condense again into a liquid but stays as a vapor. Normally the difference is not significant for most fuels used in conventional combustion systems; for hydrogen, however, the difference is about 15.6 percent. This is important since fuel cells can capture and use only the LHV of hydrogen.

7. SPE is a registered trademark owned by Hamilton Standard, a division of United Technologies Corp. Hamilton Standard was renamed Hamilton Sundstrand after United Technologies acquired the Sundstrand Corporation in 1999.

8. J. Funk and R. Reinstrom, "Energy Requirements in the Production of Hydrogen from Water," *I&EC Process Design and Development* 5 (1966): 336–342.

9. As estimated by the 2007 *Chemical Economics Handbook* Marketing Research Report for Hydrogen, published by SRI Consulting. This is the equivalent of 52.7 million U.S. tons. Of that, 12.5 million tons is produced in North America (the United States and Canada), much of it as captive commodity in a refinery or a chemical plant making ammonia, a basic component of fertilizer, where it is used without ever leaving the plant.

10. *Hydrogen & Fuel Cell Letter*, June 2007.

11. Epyx, a spinoff of Arthur D. Little, became Nuvera in early 2000 in a merger with the Italian firm De Nora.

12. *Hydrogen Tomorrow,* Report of the NASA Hydrogen Energy Systems Technology Study (Jet Propulsion Laboratory, 1975).

13. R. D. Cortright, R. R. Davda, and J. A. Dumesic, "Hydrogen from Catalytic Reforming of Biomass-derived Hydrocarbons in Liquid Water" *Nature*, August 29, 2002, 964–967.

14. *Hydrogen & Fuel Cell Letter*, June 2007.

15. Intriguingly, one of the early backers was Saudi Arabia and its then oil minister, Sheikh Ahmed Zaki Yamani.

16. Graetzel's initial project was funded by the Swiss Energy Ministry, Sandoz, and Asea Brown Boveri.

17. Yu Bai, Yiming Cao, Jing Zhang, Mingkui Wang, Renzhi Li, Peng Wang, Shaik M. Zakeeruddin, and Michael Graetzel, "High-performance Dye-sensitized Solar Cells Based on Solvent-free Electrolytes Produced from Eutectic Melts" *Nature Materials*, June 29, 2008, 626–630.

18 Michael Graetzel, "Photoelectrochemical Cells," *Nature*, November 15, 2001, 338–344.

19. *Hydrogen & Fuel Cell Letter*, April 2009.

20. Oscar Khaselev and John A. Turner, "A Monolithic Photovoltaic-Photoelectrochemical Device for Hydrogen Production via Water Splitting," *Science*, April 17, 1998, 425–427.

21. Stu Borman, "Hydrogen for Water and Light." *Chemical & Engineering News*, April 20, 1998, 11.

Chapter 5

1. *Hydrogen & Fuel Cell Letter*, January 2008; July, August 2009.

2. Wikipedia, "List of Photovoltaic Power Stations," http://en.wikipedia.org/wiki/List_of_photovoltaic_power_stations.

3. "List of Solar Thermal Power Stations," http://en.wikipedia.org/wiki/List_of_solar_thermal_power_stations.

4. With some obvious political and economic advantages, were hydrogen available in many parts of the world, oil cartels such as the OPEC of the 1970s would have a difficult time enforcing their will.

5. At the DoE's 10-megawatt Solar Two solar tower in Barstow, California, solar heat is stored at about 1,050°F (573°C) very efficiently in tanks containing molten salt. About 98 percent is the annual average round-trip efficiency. This assumes the plant runs every day dispatching energy to the grid, using solar energy collected during the day at night stored as a heat in large tanks (3 million pounds, 40 feet high, 25 feet in diameter), says James Pacheco, principal member of the technical staff at Sandia National Laboratories, which, with the National Renewable Energy Laboratory, operates Solar Two. Pacheco says heat dissipates very slowly: The tank holds about 105 megawatt-hours of electricity, and when not in use, it loses only about 100 kilowatts per hour, or about 5°F per day. "It's a very simple design, and very cheap," says Pacheco.

6. Since retiring in the 1990s, Bockris has shifted his attention to more esoteric fields, including low-temperature nuclear reactions, quantum electrochemistry, and transmutation of metals.

7. J. Bockris, *Energy: The Solar-Hydrogen Alternative* (Ultimo, New South Wales: Halstead, 1975).

8. Although no new reactors are being built in the United States, and many were canceled in the 1970s and the 1980s, the 104 reactors that were operating in the United States in 1999 were doing so more efficiently (with capacity factors approaching 80 percent in the 1990s compared to just above 50 percent in the 1970s) and were generating more of the nation's electricity (almost 25 percent, compared to about 10 percent in the 1970s) than ever before, according to the *New York Times*. A March 7, 1999, *Times* article on the twentieth anniversary of the Three Mile Island accident said that "the nuclear industry will be around for years to come, but seems to have peaked in terms of the number of working reactors and their share of power generated nationally." The last reactors (Palo Verde 1, 2, and 3, in Arizona) were ordered in October 1973, the month the Arab oil embargo began, the article said. Worldwide, 440 plants were operating as of July 1, 2011, according to the website of the World Nuclear Association; 61 were under construction, 154 were planned, and 343 plants have been proposed (http://www.world-nuclear.org/).

9. T. Dickerman (Ed.), *Renewable Energy—Experts and Advocates* (Daly City, CA: American Association for Fuel Cells, 1997), a resource book for American high school debaters.

10. Press release on Senator Webb's Web site, http://webb.senate.gov/newsroom/pressreleases/2009-11-16-01.cfm. November 16, 2009.

11. Vermont Law School "IEE Releases New Report: All Risk, No Reward for Taxpayers and Ratepayers," news release, November 13, 2009. http://www.vermontlaw.edu/Academics/Environmental_Law_Center/Institutes_and_Initiatives.htm.

12. "Korea Wins $40 Bil. UAE Nuclear Deal," *Korea Times*, December 27, 2009.

13. ITER Facts & Figures Web site,. http://www.iter.org/factsfigures.

14. "Here Comes the Sun," *New Statesman*, November 26, 2009, www.newstatesman.com/environment/2009/11/fusion-reactor-iter-energy; "Hot, Medium and Cool Fusion," MSNBC, Cosmic Log, December 3, 2009. MSNBC.com science editor Alan Boyle reports that in addition to ITER's Tokomak magnetic fusion technology, other unorthodox technologies such as laser fusion and U.S. Navy–backed inertial electrostatic fusion are being investigated in the United States with much smaller budgets and typically on a smaller scale; http://cosmiclog.msnbc.msn.com/_nv/more/section/archive?year=2009&month=12&ct=a&pc=25&sp=25#December 2009 archive;Science Insider, "ITER Fusion Reactor Faces New Delay," November 19, 2009. http://news.sciencemag.org/scienceinsider/2009/11/iter-fusion-rea.html.

15. *Environmental Leader,* December 21, 2009, http://www.environmental leader.com/2009/12/21/citing-market-conditions-bp-solar-dismantles-factory-addition/.

16. *Hydrogen & Fuel Cell Letter,* January 2010.

17. *Hydrogen & Fuel Cell Letter,* April 2009.

18. Air Quality Management District Media Office e-mail communication to the author, October 19, 2010.

19. "Running out of Juice," *Economist,* January 29, 2010.

20. http://h2fcvworkshop.its.ucdavis.edu.

21. Linda Church Ciocci, "Hydropower, The Nation's Leading Renewable Energy," in Dickerman, *Renewable Energy.*

22. Energy in Iceland, *Iceland Trade Directory;* undated. http://www.icelandexport .is/icelandexport2/english/industry_sectors_in_iceland/Energy_in_Iceland/.

23. *Hydrogen & Fuel Cell Letter,* June 1998.

24. "Downturn Ends Boom in Solar and Wind Power," *New York Times,* February 4, 2009.

25. David Kearney of Kearney & Associates, a former Luz vice president.

26. Solar Photovoltaic Industry—Looking through the Storm, January 21, 2009. www.dbcca.com/dbcca/EN/_media/Global_Markets_Steve_ORourke_Solar _January_2009.pdf.

27. Photovoltaics, *Wikipedia,* page of January 16, 2010. http://en.wikipedia.org/ wiki/Photovoltaics.

28. "UD-Led Team Sets Solar Cell Record, Joins DuPont on $100 Million Project," *University of Delaware Daily,* July 23, 2007; http://www.udel.edu/PR/ UDaily/2008/jul/solar072307.html.

29. "Sharp Develops Solar Cell with World's Highest Conversion Efficiency of 35.8%," *physorg.com,* October 22, 2009, http://www.physorg.com/ news175452895.html.

30. Solar Electricity Prices, June 2011. http://www.solarbuzz.com/facts-and-figures/retail-price-environment/solar-electricity-prices.

31. Ralph Overend and Susan Moon, Biomass, in *Solar Energy,* Maureen McIntyre (ed.),(Boulder, Colorado: American Solar Energy Society, 1997).

32. Ibid.

33. Ibid.

34. John Reilly and Sergey Paltsev, *Biomass Energy and Competition for Land,* Report No. 145, MIT Joint Program on the Science and Policy of Global Change, April 2007).

35. *Hydrogen & Fuel Cell Letter,* November 1996.

36. *Hydrogen & Fuel Cell Letter,* September 2002; March 2007.

37. "DOE Beaming in on Solar Satellites," *Engineering News Record,* March 2, 1978, 12.

38. "Energie aus dem Weltall mit der 'Kraftsoletta,'" *Handelsblatt*, September 6, 1977.

39. There are two basic types: closed cycle and open cycle. In a closed-cycle system, warm surface seawater and cold deep seawater are used to vaporize and condense the working fluid such as ammonia, which then drives the turbine generator in a closed loop. In an open-cycle system, surface seawater is flash-evaporated in a vacuum chamber, and the resulting low-pressure steam drives a turbine generator. Cold seawater is then used to condense the steam after it has passed through the turbine. The open cycle therefore can be configured to produce fresh water as well as electricity.

40. Vega does not consider mere paper studies "as activity, given the numerous studies, in at least seven different languages, in the last century." Fax to the author, November 12, 1997.

41. L. A. Vega, "Ocean Thermal Energy Conversion Primer," *Marine Technology Society Journal* 6, no. 4 (Winter 2002/2003):, 25–35.

42. *Strategic Plan for the Geothermal Energy Program* (Washington, D.C.: U.S. Department of Energy, Office of Geothermal Technologies, June 1998).

43. There are two main types of geothermal power plants: flash steam type and binary cycle. The flash steam plant is driven by pressurized, hot (300–700°F) steam brought up from depth of as much as more than 10,000 feet. When this pressure is reduced at the surface by about a third or so, the water "flashes"—explosively boils—into steam, which drives the turbine and generator. Binary plants are driven by lower-temperature (212–300°F) geothermal fluids. Here, the fluids are passed through a heat exchanger that heats a secondary, usually organic, working fluid such as isopentane, which turns into steam at temperatures lower than water. That organic steam drives the turbine and the generator before being recondensed into fluid and vaporized again by geothermal heat.

Chapter 6

1. *Hydrogen & Fuel Cell Letter*, July 2008.

2. *Hydrogen & Fuel Cell Letter*, July 2011.

3. *Hydrogen & Fuel Cell Letter*, February 1996; October 1999; April, September 2002; January 2010.

4. Guy Procter, "World First Ride: Hydrogen," *Motorcycle News*, February 10, 2010.

5. *Hydrogen & Fuel Cell Letter*, April 2008.

6. *Hydrogen & Fuel Cell Letter*, October 2009.

7. "Can Motor City Come Up with a Clean Machine?" *New York Times*, May 19, 1999.

8. There are other types, such as chemical hydrides and liquid organic hydrides.

9. E. Dickson, T. Logothetti, J. Ryan, and L. Weisbecker, "The Use of Personal Vehicles within the Hydrogen Energy Economy—An Assessment," in *Proceedings of Hydrogen Economy Miami (THEME) Conference* (1974).

10. U.S. Department of Health and Human Services, Public Health Service, Agency for Toxic Substances and Disease Registry, *Methanol Toxicity* (April 1997).

11. *Hydrogen & Fuel Cell Letter*, July 2003, November 2008.

12. "CarbonNanotubes by the Metric Ton," *Chemical and Engineering News*, November 12, 2007.

13. "Selling Fuel Cells," *Economist*, May 25, 1996.

14. *Hydrogen & Fuel Cell Letter*, February 1999.

15. *Hydrogen & Fuel Cell Letter*, August 2009.

16. *Hydrogen & Fuel Cell Letter*, June 2004.

17. The Bavarian government, which is trying to encourage hydrogen technology as a high-payoff future technology within its borders, assigned coordinating responsibility for the project to Ludwig-Boelkow-Systemtechnik, a subsidiary of the Ludwig-Boelkow Foundation near Munich, which has been active in promoting and supporting hydrogen energy technologies since the early 1980s.

18. Gasoline-powered golf carts were banned in California in 1997.

19. *"Alternative Energy Sources for Road Transport—Hydrogen Drive Test"* in German. (Cologne: Verlag TÜV Rheinland for the German Federal Ministry for Research and Technology, 1990.

20. In the more sophisticated internal mixture formation, hydrogen is injected directly into the combustion chamber during combustion, as in a diesel engine. For this, hydrogen must be supplied at high pressure. This approach eliminates backfire and self-ignition in the compression phase, and there is no loss of volumetric efficiency. Engine operation is characterized by high power density and low fuel consumption, according to the TUV report, but the system is technically more complicated and problems may occur with the mixture homogenizing at higher engine speeds. It is the approach favored by BMW, which from the beginning of its work on hydrogen internal combustion engines has concentrated on the use of liquid hydrogen.

21. "Alternative Energy Sources for Road Transport."

22. Peschka and BMW parted company shortly after the completion of their first vehicle.

23. The Perris Smogless Automobile Association ceased work in 1973. Paul Dieges maintained his involvement with hydrogen through the late 1990s, working with the southern California chapter of the American Hydrogen Association.

24. Perris Smogless Automobile Association, *First Annual Report* (Perris, California: 1971).

25. Walter Peschka, *Liquid Hydrogen—Fuel of the Future* (New York: Springer-Verlag, 1992).

Chapter 7

1. Hydrogen-fueled locomotive unveiled. *The Orange County Register*, January 29, 2010.

2. Governor Schwarzenegger website, Speeches, January 28, 2010 (http://gov.ca.gov/speech/14303/).

3. *Hydrogen & Fuel Cell Letter*, August 2009; February 2010.

4. *Hydrogen & Fuel Cell Letter*, February 2009.

5. *Hydrogen & Fuel Cell Letter*, July 2010.

6. A. Appleby and F. Foulkes, *Fuel Cell Handbook* (New York: Van Nostrand Reinhold, 1989; Krieger, 1993). This book, one of the standard reference texts on fuel cell technology, provided much of the historical and technical information for this section. Another main source, especially for the early history of fuel cell technology, is *Fuel Cell Systems*, ed. L. Blomen and M. Mugerwa (New York: Plenum, 1993).

7. Karl Kordesch and Guenter Simader, *Fuel Cells and Their Applications* (Weinheim, Germany: VCH Verlagsgesellschaft mbH, 1996).

8. Ibid. Hydrazine, ammonia, and methanol have been investigated as "carriers" of hydrogen for use in fuel cells, in part because hydrogen is difficult to store onboard. Of the three, hydrazine, a liquid, was found to be the most promising for fuel cell use because it is easily dissolved in aqueous electrolytes—the electrochemical heart of the fuel cell—and because it is readily oxidized. However, its potential use for normal ground transportation is unlikely because it is expensive to manufacture: Appleby says the cost is about fifteen to twenty times as high as for hydrogen for the same amount of energy. But even if cost were not a factor, its suitability for normal road transport with its everyday hazards ended when it was found that the stuff is highly toxic—human tolerance is about 1 ppm in air—and that it poses a severe explosion hazard when exposed to heat or reacts with oxidizing materials. Methanol, despite some biomedical and environmental hazards, is currently the fuel of choice. Ammonia, the least reactive of the three, also has good points in its favor. It is available worldwide (as fertilizer, for instance), costs are only slightly higher than those of methanol on a Btu basis, and it is easy to handle under low pressure. It has low toxicity, it is easy to handle as a pure fuel in lightweight containers—and leaks would be easily detected because of its pungent smell. It has not received much attention as a power source for fuel cells except for remote or specialized application, but some speculate this may change with future vehicle applications—an ammonia-fueled tractor for farmers?

9. "At Last, the Fuel Cell" (article) and "The Third Age of Fuel" (editorial). *The Economist*, October 23, 1997.

10. Ulf Bossel, *The Birth of the Fuel Cell 1835-1845* (Oberrohrdorf, Switzerland: European Fuel Cell Forum, 2000); *Hydrogen & Fuel Cell Letter*, September 2000.

11. UTC Power press release, April 15, 2010.

12. Even a cursory description of the many programs that got underway in those years would require much too much space here. For details, interested readers should see the books by Appleby and Foulkes, *Fuel Cell Handbook*; Blomen and Mugerwa, *Fuel Cell Systems*; and Kordesch and Simader, *Fuel Cells and Their Applications*.

13. Tokyo Electric Power Company (principal investigator: K. Shibata), "Demonstration Testing of 11 MW Phosphoric Acid Fuel Cell Power Plant—from Planning to Power Generation" (Palo Alto, CA, 1992).

14. E-mail from Akifusa Hagiwara, manager of the Material Science Group, Energy and Environment R&D Center, Tokyo Electric Power Co., October 12, 1999.

15. Roger Snodgrass, "Solar Tech Company Helps Catch a Brighter Ray," *Los Alamos Monitor*, January 14, 2010.

16. *Hydrogen & Fuel Cell Letter*, November 2009.

17. Scott Kirsner, "While World-changing Technology May Take Time to Develop, Patience Can Pay off for Investors," *Boston Globe*, February 14, 2010.

18. Because they are very quiet and are inherently more efficient than diesel or other internal combustion power plants, fuel cells have long been regarded by the navies of the United States, Germany, Canada, Australia, and Sweden as an attractive means of propelling small submersibles (such as unmanned underwater vehicles used for marine research, pipeline repair, salvage, and exploration), and even full-size submarines.

19. In addition, Kordesch and Simader differentiate among different fuel cell systems: direct fuel cells (those that electrochemically convert fuels such as hydrogen, but also organic or nitrogenous compounds such as ammonia and hydrazine, even metals and hydrogen/halogen combinations at low, intermediate or high temperatures); indirect fuel cells (the type that requires a reformer to extract hydrogen, or a fuel cell running on biochemical fuels that are decomposed via enzymes to produce hydrogen); and regenerative fuel cells (systems that can work either as a fuel cell–producing electricity, or as a reversible system—as an electrolyzer—in which electrical energy can be stored by splitting water, for example, into hydrogen and oxygen).

20. Linc Energy, press release, June 29, 2010.

21. Also known as polymer electrolyte fuel cell (PEFC). Another term occasionally used is *ion exchange membrane* (IEM).

22. Christoph Wannek, "High-Temperature PEM Fuel Cells: Electrolytes, Cells, and Stacks" in *Hydrogen and Fuel Cells: Fundamentals, Technologies and Applications*, ed. Detlef Stolten (Weinheim : Wiley-VCH). Stolten is director of the Institute for Fuel Cell Research at the Juelich Research Center in Germany, and he was conference chairman.

23. *Hydrogen &Fuel Cell Letter*, April 1997.

24. *The Hydrogen Letter*, November 1993

25. *Hydrogen & Fuel Cell Letter*, March 2010; Internet video with Sridhar presentation is on the *Bloom Energy website* at http://bloomenergy.com/bloom-energy-launch-event-staff/.

26. *Hydrogen & Fuel Cell Letter*, March 2010.

Chapter 8

1. Tom Koehler, "A Green Machine," *Boeing Frontiers*, May 2008.

2. *Hydrogen & Fuel Cell Letter*, April 2007; May 2008; August 2009; June 2010.

3. *Hydrogen & Fuel Cell Letter*, September 2000.

4. *Fuel Cell Today,* www.fuelcelltoday.com/events/industry-review.

5. The 1960s TU-154, workhorse of East Bloc aviation, is still in service today in various parts of the world. Russia's national Aeroflot carrier phased out the last one in 2009.

6. *The Hydrogen Letter*, Special, April 1988.

7. Personal communication (phone interview). Years later it emerged that the Russians had in fact developed a system for pumping liquid hydrogen. Daimler-Benz Aerospace, which in the mid-1990s embarked on a project of demonstrating a liquid hydrogen–powered commuter plane, said in a brochure that the new plane's tank pump would be based on the TU-155's pump, but that it would be redesigned because the old Russian pump had produced too much heat, resulting in too much boil-off of fuel.

8. "Liquid Hydrogen Fueled Aircraft," NASA Symposium (Langley Research Center: May 15–16, 1973). As a fascinating sidelight, Boeing presented a concept sketch of a liquid hydrogen-fueled Boeing 747 at the symposium.

9. The engine of Conrad's plane was essentially standard, though its fuel injection system had been modified (with help from G. Daniel Brewer) to handle liquid hydrogen. The timing and control systems for actuating the injection valves were devised by Roy Parsons, an electronics design engineer from Pompano Beach, Florida. The tank had been fabricated and donated by Consolidated Precision Corporation of Riviera Beach. The flight was preceded by several months of ground testing.

10. Stuart F. Brown, "Inside the Skunk Works," *Popular Science*, October 1994, 52–60.

11. A paper by Brewer describing the various designs was published in the first edition of the *International Journal of Hydrogen Energy* in early 1976: G. D. Brewer, "Aviation Usage of Liquid Hydrogen Fuel—Prospects and Problems," *International Journal of Hydrogen Energy* 1, no. 1 (1976): 65–88.

12. The reasons for the difference in tank arrangements have to do with the fuels' characteristics. JP-4 is stable at ambient temperatures, and the shape of the tank therefore does not make much difference, permitting JP-4 to be stored in the wings, which also helps to stabilize and strengthen the wing structure.

Supercold liquid hydrogen, on the other hand, boils off, creating pressure gas. To contain this pressure and minimize odd angles and curves where pressure might build to the bursting point, a round, or nearly round, perhaps elliptical, pressure tank is best.

13. This works for both hydrocarbon fuels and hydrogen, but hydrogen is especially suitable because it mixes more easily and therefore a simpler combustor design could be used.

14. See note 7.

15. G. D. Brewer, "Plan for Active Development of LH2 for use in Aircraft," *International Journal of Hydrogen Energy* 4(1979):, 169–177.

16. As the Bonn correspondent for McGraw-Hill World News and for *Business Week* at the time, I reported the story.

17. A good source of information on these early efforts is Russell Hannigan's *Spaceflight in the Era of Aero-Space Planes* (Malabar, FL: Krieger, 1994).

18. It is not likely that any NASP-type transatmospheric craft was ever built. However, the possible secret existence of a very fast (Mach 4.5–6), very high-flying aircraft powered by liquid hydrogen, liquid methane, or some other unconventional fuel has excited the imagination of enthusiasts and professional analysts. There has been speculation that this plane, generally referred to as Aurora, is a successor to the supersonic SR-71 reconnaissance plane. The U.S. government has steadfastly denied its existence. The Federation of American Scientists' aerospace expert, John Pike, at one point devotes almost a dozen pages to "Aurora/Senior Citizen" on the "Mystery Aircraft" section of the FAS Intelligence Resource Program Web (http://www.fas.org/irp/mystery/aurora.htm)site. He cautiously noted that "although there is a growing body of evidence that could be interpreted to suggest the existence of one or more advanced aircraft behind the veil of government secrecy, the evidence remains suggestive rather than conclusive." There have not been any reported sightings or other developments for several years, Pike noted in a fall 1999 e-mail message, adding that "the trail has gone cold." However, the respected aviation writer Bill Sweetman wrote at the end of a very long 2006 *Popular Science* article, "The Top-Secret Warplanes of Area 51," that he believed Aurora is for real, although it may not be fueled by hydrogen or methane, as originally suggested twenty to twenty-five years earlier (http://www.popsci.com/military-aviation-space/article/2006-10/top-secret-warplanes-area-51#). He concluded, "Years of pursuit have led me to believe that, yes, Aurora is most likely in active development, spurred on by recent advances that have allowed technology to catch up with the ambition that launched the program a generation ago."

19. *Hydrogen & Fuel Cell Letter*, June 2011.

Chapter 9

1. *Hydrogen & Fuel Cell Letter*, April 2010.

2. http://en.wikipedia.org/wiki/Yin_and_yang.

3. Robert B. Rosenberg, and Esther R. Kweller, "Catalytic Combustion of Reformed Natural Gas," *Appliance Engineer* 4 (August 1970): 32–36.

4. *The Hydrogen Letter*, July 1990.

5. According to a report given at the 1996 World Hydrogen Energy Conference, Fraunhofer Institute scientists working on this project also developed several types of hydrogen burners with very low nitrogen oxide emissions, including a diffusion-type burner for retrofitting conventional natural gas–type burners. They also fabricated and tested a 33- kilowatt gas absorption chiller and a 21-kilowatt space heater.

6. Jon B. Pangborn, Maurice I. Scott, and John C. Sharer, "Technical Prospects for Commercial and Residential Distribution and Utilization of Hydrogen," *International Journal of Hydrogen Energy* 2 (1977): 431–445.

7. Siobhan Wagner, "Hydrogen Embrittlement Could Lead to Failures of Fuel-Cell Cars," *Engineer*, www.*theengineer.co.uk*, August 4, 2010.

8. *Hydrogen & Fuel Cell Letter*, June 2010.

9. *Hydrogen & Fuel Cell Letter*, July 2009.

10. *Hydrogen & Fuel Cell Letter*, May 2009, July 2009.

11. KBB Underground Technologies, "Grid Scale Energy Storage Based on Pumped Hydro, Compressed Air and Hydrogen," June 15, 2010; http://catedras empresa.esi.us.es/endesared/documentos/jornada_almacenamiento/Fritz _Crotogino.pdf; KBB Underground Technologies, "Druckluftspeicher—Ein Element zur Netzintegration Erneuerbarer Energien," June 18, 2009.

12. H. T. Everett Jr., NASA Fluids Management, e-mail to the author, August 19, 2010.

13. G. Kaske, P. Schmidt, and K.-W. Kannengiesser, "Vergleich zwischen Hochspannungsgleichstromübertragung und Wasserstofftransport" (Comparison between High-Voltage DC Transmission and Hydrogen Transport), VDI Energy Technology Society Meeting, Stuttgart, 1989.

Chapter 10

1. *Hydrogen & Fuel Cell Letter*, June 2007, January 2010.

2. Wikipedia, "Trans Fat," *http://en.wikipedia.org/wiki/Trans_fat*, updated August 8, 2010; Mayo Clinic, updated July 8, 2010, "Trans Fat Is Double Trouble for Your Heart Health," www.mayoclinic.com/health/trans-fat/CL00032.

3. Tokiaki Tanaka, "Hydrogen Economy from the Viewpoint of Nonferrous Extractive Metallurgy," *Journal of Metals* (December 1975): 6–15.

4. Lutero Carmo de Lima, Joao Batista Furlan Duarte, and T. Nejat Veziroglu, "A proposal of an Alternative Route for the Reduction of Iron Ore in the Eastern Amazonia," *International Journal of Hydrogen Energy* 29 (May 2004): 659–661.

5. J. Gretz, W. Korf, and H. Lyons, Wasserstoff in der Stahlindustrie. *Das Solarzeitalter* 4, (1990): 20–22.

6. M. L. Yaffee, "Atomic Hydrogen Rocket Fuel Studied," *Aviation Week & Space Technology,* November 25, 1974, 47.

7. Robert Hazen, *The Alchemists: Breaking Through the Barriers of High Pressure* (New York: Times Books, 1993).

8. Ibid.

9. "Big Gun Makes Hydrogen into a Metal," *New York Times,* March 26, 1996. A more detailed account by William Nellis is the May 2000 cover story of *Scientific American,* "Metallic Hydrogen—The Stuff of Jupiter's Core Might Fuel Fusion Reactors."

10. Hazen, *The Alchemists.*

11. Bryan Palaszewski, "Atomic Hydrogen Propellants: Historical Perspectives and Future Possibilities," NASA Scientific and Technical Information Program Office, 1993.

12. "The thrust produced per unit rate of consumption of the propellant usu. specified in pounds of thrust per pound of propellant used per second and forming a measure of the efficiency of performance of a rocket engine" (*Webster's Third New International Dictionary*).

13. Bryan Palaszewski, "Solid Hydrogen Experiments for Atomic Propellants: Particle Formation Energy and Imaging Analyses," NASA Glenn Research Center, 2002.

14. Bryan Palaszewski, "Solid Hydrogen Experiments for Atomic Propellants: Particle Formation, Imaging, Observations, and Analyses,". Glenn Research Center, 2005.

15. William P. Fife, "The Medical Applications of Hyperbaric Molecular Hydrogen Phase II," (unpublished paper draft, July 1997).

16. According to William Fife, Lavoisier, one of the discoverers of hydrogen, discussed the possible use of hydrox in one of his seminal studies, published by France's Academy of Sciences in 1789.

17. H. G. Schlegel, and R. M. Lafferty, "The Production of Biomass from Hydrogen and Carbon Dioxide," *Advances in Biochemical Engineering. Vol. 1* (Heidelberg and New York: Springer, 1971, 143–167).

18. Sharon Begley, "ET, Phone Us," *Newsweek,* October 12, 1992. The definition of the "water hole" is actually a bit wider. The Oak Ridge Observatory in Harvard, Massachusetts, tunes in to frequencies in the range from 1,400 to 1,720 megahertz, a spectrum that takes the Doppler effect into account.

19. Ibid.

20. Malcolm W. Browne, "Big Gun Makes Hydrogen Into a Metal," *New York Times,* March 26, 1996.

Chapter 11

1. Also, a much smaller portion of the total fuel carried by both planes would presumably have been spilled.

2. Malcolm W. Browne, "Hydrogen May Not Have Caused Hindenburg's Fiery End," *New York Times*, May 6, 1997; Richard G. Van Treuren, "Odorless, Colorless, Blameless," *Air & Space*, April/May 1997, 14–16.

3. Mariette DiChristina, "What Really Downed the Hindenburg," *Popular Science*, May 1997, 70–76.

4. A. A. Du Pont, "Liquid Hydrogen as a Supersonic Transport Fuel," *Advances in Cryogenic Engineering* 12: 1–10, 1967.

5. Brewer based his conclusions on the report by the Spanish Ministry of Transport and Communications that said, among other things, that the KLM plane's fuselage was not particularly deformed, and neither the impact against the Pan Am plane nor the one against the ground was particularly violent. Brewer thought that under these circumstances, any hypothetical liquid-hydrogen tanks, which in some Lockheed designs of those years would have been located inside the fuselage in two tanks, one forward and one aft of the passenger compartment rather than in the wings as in standard, commercial kerosene-fueled B-747s, would probably not have been ruptured—or at least not both of them. The only fire would have been due to the relatively small amounts of hydrogen in the feed lines running through the wings to the engines.

6. However, even a weak spark, such as one caused by static electricity from a human body, is enough to ignite any of these fuels. Such a spark produces about 10,000 microjoules.

7. Michael R. Swain, "Fuel Leak Simulation," Paper and video presentation, Twelfth Annual U.S. Hydrogen Meeting, Washington D.C., March 6–8, 2001.

Chapter 12

1. E-mail communications from Michael Oppenheimer, Department of Geosciences, Princeton University, and Daniel Sperling, Institute of Transportation Studies at the University of California, Davis, September 29, 2010.

2. Three reports from the National Research Council cited in this paragraph: "Review of the Research Program of the FreedomCAR and Fuel Partnership," second report, 2008; "America's Energy Future: Technology and Transformation," 2009; "Review of the Research Program of the FreedomCAR and Fuel Partnership,"; third report, 2010.

3. *Hydrogen & Fuel Cell Letter*, August 2009.

4. For example, in 1976 Marchetti proposed Energy Islands, truly gargantuan nuclear energy facilities hidden away in remote parts of the Pacific. As a concrete example, he suggested installing a terawatt's worth of nuclear power—five barge-mounted reactors of 200 gigawatts each—in Canton Island, a 6- by 9-mile atoll in the equatorial boondocks of Micronesia, some 1,300 miles west of New Guinea. (A terawatt is about 769 times the electrical energy produced by the largest 1,300 megawatt electric nuclear reactors built so far, somewhere between one-seventh and one-tenth of the total power capacity of the planet.) The energy produced would be shipped as liquid hydrogen by behemoth 1,600-foot-long

barges to population centers. Marchetti was not fazed by charges of technological gigantism: "The size of such reactors is certainly mindboggling to nuclear engineers, as would be the sight of a 1,000 megawatt generator to Thomas Edison for whom a giant generator was in the range of hundreds of kilowatts," he said at the time in a paper. He repeated his main concepts in a keynote speech at the 1998 World Hydrogen Energy Conference in Buenos Aires. According to Marchetti, a central rationale for locating giant hydrogen production centers far from civilization would be the superior transportability of gaseous fuels: electric power stations, he said, can "see" about 100 kilometers—the mean distance of electricity transport, given power consumption density patterns in industrialized nations. For a gas system with pipelines, it is about 1,000 kilometers—an area 100 times larger. Asked by an audience member whether this does not fly in the face of current trends such as energy decentralization, Marchetti replied, "Small may be beautiful, but big is cheap after the technology has had time to settle."

5. C. Marchetti, "Primary Energy Substitution Models," staff paper (based on a 1974 Moscow lecture) International Institute for Applied Systems Analysis, Schloss Laxenburg, Austria, 1975.

6. Marchetti sums up these rules as follows: "The fractional rate at which a new commodity penetrates a market is proportional to the fraction of the market not yet covered. It includes two constants as characteristics of the particular commodity and market."

7. Arnulf Grübler and Nebojša Nakićenović , "Decarbonizing the Global Energy System" (Laxenburg, Austria: International Institute for Applied Systems Analysis, 1997).

8. Jeremy Rifkin, *The Hydrogen Economy* (New York: P. Tarcher/Putnam, 2002).

9. William J. Mitchell, Christopher E. Borroni-Bird, and Lawrence D. Burns, *Reinventing the Automobile: Personal Urban Mobility for the 21st Century* (Cambridge, MA: MIT Press, 2010).

10. John Hofmeister, *Why We Hate the Oil Companies: Straight Talk from an Energy Insider* (London: Palgrave Macmillan, 2010).

Index